Teaching Physical Education
A Systems Approach

Teaching Physical Education

A Systems Approach

Robert N. Singer
The Florida State University

Walter Dick
The Florida State University

Houghton Mifflin Company • Boston

Atlanta
Dallas
Geneva, Illinois
Hopewell, New Jersey
Palo Alto
London

Photo on p. 19, Brown Brothers; p. 70, Sports Equipment, Inc.; pp. 282,
284, Wide World, Inc.; other photos by Frank Siteman. Thanks to Beaver
Country Day School, Medford High School, Melrose High School, Rox-
bury Latin School, Tufts University, and Winchester High School, all of
Massachusetts.

Contents

v

Preface

This book aims to present the student, contemplating a teaching career in physical education, with an exciting approach currently being tried by many other educational disciplines. The systems approach is both theoretical (in its formulation of models) and practical (in its application of research to real situations). It's our goal to convince the prospective teacher of the merits of a teaching approach which is logical, conceptual, and practical. Though the systems approach is surely no easy route for the teacher—it requires great thought in the preparation of courses and curriculum—it does bring in-depth analysis to bear with potential payoffs far exceeding those currently being realized.

In all quarters of education, increasingly dramatic innovations are being introduced. The student as an individual is of prime concern. Physical educators are typically concerned with psychomotor behaviors (skill acquisition, meaningful movement, physical fitness), cognitive behaviors (understanding rules, strategies), social behaviors (interpersonal relationships), and affective behaviors (values, feelings). To effectively modify such a variety of behaviors is a challenge to the teacher. How can the teacher accomodate individual differences while determining behavior expectations? The answer may be the learning systems approach.

An assessment of student interest, characteristics and skills, available resources, necessary conditions and materials, and terminal objectives are primary components in the systems approach to learning. Media and instructional technology shape instructional strategies. Goals and behavioral objectives are specified and evaluated to determine if they have been realized. A course, a unit, a program, and a curriculum are all tied together in a systematic manner. The systems approach requires the teacher to analyze all aspects of the learning process in a more concrete fashion.

The loose structure of educational experience becomes more meaningful when the educator is required to examine carefully both his starting point and what he is attempting to accomplish. Students have complained for years about a lack of concern by teachers for individual differences (and the need for specification of desirable behavioral outcomes). More beneficial learning climates can and must be arranged. Systematic models enable both educators and students to proceed with confidence instead of confusion and instability.

Systems models were first devised by industry to accomplish a more effective return for time and dollars spent. Educators have since borrowed, adapted, and applied concepts from the systems approach to their own situations, apparently with successful results. These efforts, occurring only in the last few years, are most noticeable in the cognitive domain; the possible applications in the psychomotor domain have remained virtually unexplored. Specifically, physical educators have failed to implement such procedures in their teaching practices.

Yet, the handwriting on the wall is clear; students are not satisfied with authoritarian, formal, and insensitive techniques of instruction. They want to enjoy and benefit from their experiences. They want to know precisely what is expected of them. They want to achieve goals that are primarily student centered, not group centered. If these needs are to be realistically met, teaching procedures must be modified accordingly. The teacher must be both knowledgeable and flexible.

The demand for accountability in teaching has proved to be a challenge to educators everywhere. Teachers and administrators at all levels of education recognize the consequences of not attempting to respond to criticism of poor and mediocre instruction, inadequate curricula and programs, and insufficient consideration for student needs. Though there is a long way to go, the future looks brighter than it has in some time. Why? Because advances in educational technology, learning principles and educational theory, and instructional strategies are being combined with progressive, sensitive, and creative leadership. The outcome—more acceptable and palatable teaching procedures—is refreshing and exciting.

The systems approach should be an option in any consideration of teaching methods in physical education. We hope that you as a teacher-to-be will accept the approach suggested and try to implement it in your own teaching situation. It is an option both you and your future students are waiting for.

Walter Dick
Robert N. Singer

PART ONE
Introduction

The first five chapters of this book deal with the nature of the systems approach as applied to instruction. Chapters One through Four explain why this approach is a sound basis for improving teaching. The approach is based on the latest technological, scientific, and educational developments. It also makes sense.

We all advocate high-quality education. But what is high-quality education? Once worthwhile instructional goals are established, the procedures used to reach them should be appropriate, effective, and efficient. There is no one right way to do anything, but a structure must be established to insure that all pertinent factors are considered and instruction proceeds in an organized and effective way. The systems approach is presented to satisfy this requirement. It can be used in any instructional situation, but in this book is designed for the physical education teacher.

Chapter One provides an overview of current issues bearing on all areas of education, and a perspective for viewing directions in physical education. Chapter Two describes the principal traditional and creative approaches to teaching physical education. Both chapters urge sound teaching methodologies to produce realistic, specified, and measurable outcomes. Chapter Three is an overview of the systems approach to instruction detailed in later chapters.

A variety of systems approach models are used today in industry, the military, and all areas of education. The authors believe their model to contain valid considerations in preparing and delivering instructional content. Chapter Four describes the use of one systems approach and offers convincing data on its effectiveness.

Chapter Five, the last in Part One, analyzes four categories of behaviors. It helps make it easier to logically apply the systems approach. Part Two contains a detailed description of each component of the systems approach model. Part Three applies the systems approach to a variety of instructional settings.

1

Physical Education and the Educational Process

Educational institutions, policies, and programs are under fire from the left and from the right, from the inside and from the outside. Controversies center around curriculum, instructional processes, student differences, and student outcomes. Physical educators face the same issues; they must confront and resolve the same nagging problems, acknowledged or otherwise, whether in gymnasium, pool, or classroom.

How can teachers most effectively challenge students to reach the goals of a course or program? Using today's advanced technology, how can teacher and media best interact to foster favorable learning climates? How can individual students realize their potential in a group setting? To what degree can teachers first specify, then evaluate their objectives in order to give an account of themselves to administrators, students, and the concerned public? These are a few of the intriguing issues presently facing all educators. To be accountable, educators must approach their tasks in a more systematic, planned way than in the past. To make meaningful changes in physical education, we must understand the events of the past, determine where we are now, then consider where we are headed.

Chapter Contents

3

Student Objectives

The student who understands this chapter should, with the book closed, be able to:

1. Identify four new educational concepts that are emerging as a result of present dissatisfaction with curricula and instructional techniques
2. Describe at least four objectives of physical education as set forth by the American Association for Health, Physical Education, and Recreation
3. Identify five stages in the historical development of physical education
4. Suggest at least three educational concepts that oppose the traditional practice of prescribing one physical education program without regard for student differences

A Perspective

Many have complained for years about the inadequacies of formal education. At times educators have responded with considerable sensitivity by modifying programs; at other times they have showed unconcern or even hostility. As in the past, contemporary educators enjoy little peace. Experimental programs, new curricula, new courses, and new experiences are being tried at all levels in an attempt to right the educational process.

Very few people—from legislators to blue collar workers and including students, teachers, and administrators—are satisfied with education in the United States today. Some want it more flexible, others more traditional. Some say, "Let's return to the good old days," while others desire change. Criticisms from all quarters are becoming more vocal. Ideas about educational programs, practices, and theory are no longer the exclusive prerogative of educators, for students and an enlightened public are becoming increasingly involved. Like it or not, teachers and administrators are being pressed to show that they can "deliver the goods."

At one time educators were allowed or even encouraged to be critical of educational practices. Professionals were to identify strengths and remedy weaknesses in the structure, it was said, but it was "hands off" to outsiders. Students in particular were not to "meddle." But the professionals have come up with solutions satisfactory to none.

Yet not all that has occurred, past or present, is all bad; education has made some progress in meeting the changing needs of society. With more and more demands burdening educators from an increased number of sources, it is unrealistic to expect an educational panacea. To satisfy diverse needs, flexibility in educational programming is necessary. Educators, sensing the need to respect individual differences in rates of learning, have developed individual and small-group approaches to influencing student behaviors. Instructional techniques compatible with individual learner characteristics are proposed. Despite these improvements, an air of uneasiness still lingers around the educational process.

As the physical educator of tomorrow, you may find little comfort in this atmosphere. Does it threaten you? Perhaps you are tempted to leave education and embark on a different career. Don't leave; the changes in the air should please and excite you. You should be pleased to be a part of the many exciting present and future changes in the educational scene. You have reason to hope that the things you disliked in elementary, middle, and high school will be remedied; no doubt you will offer more meaningful experiences to your students than were offered to you. If you like challenges, and those of us who compete in sport do, then the teaching situation will be most satisfying.

The typical image of the authoritarian teacher administering exercises to a class in a rigidly prescribed manner should probably be cast aside. Physical education often suffers from a controlled, stifling environment, where the teacher dominates the scene. The situation in which students are expected to meet normative standards or suffer punishing exercises must be replaced by one in which the students are individually challenged, stimulated, and fulfilled. The photograph on page 6 tells the story. The changing of this situation is not easy but nevertheless imperative. The faces we see in the photograph on page 7 reflect a more desirable atmosphere. The physical education atmosphere should be more relaxed. The play should be spontaneous, enjoyable, fulfilling. After all, one of the leading objectives in physical education is to encourage the development of favorable attitudes toward physical activity, thereby promoting participation outside of class and when formal schooling is completed.

This is only one example of the changing scene. As you read further, you will see other examples of possible changes designed to serve the long-neglected needs and interests of the students. Ridiculing current practices is easy, but suggesting alternatives to tradition is brave, provoking, and often leads to resistance from traditionalists. Though everyone complains about problems in physical education programs but does little to remedy them, you have the opportunity to innovate. We hope this book will provide directions for change and positive outcomes in physical education programs.

The Changing Educational Scene

Many problems in physical education are quite dissimilar to those faced by other fields of education. We have some unique problems. But the really important issues can be found throughout education. A philosophical viewpoint on directions in education

"Physical education often suffers from a controlled, stifling environment, where the teacher dominates the scene."

involves all of its components. It is therefore necessary to establish a point of view toward education and its purpose and means of functioning before we worry about very specific problems. Here is one example of a concept of the way education should proceed.

The Role of Media

George Leonard's book, *Education and Ecstasy*, contains many provocative ideas, observations, and suggestions for change. His basic premise is that "learning involves interaction between the learner and his environment and its effectiveness relates to the frequency, variety, and intensity of the interaction (p. 19)." He suggests that the *learning environment* is more important than ever before. Teachers should consider different forms of stimulation, of teaching media. No longer does education occur in one environment, such as the class lecture. Rather, *students can be stimulated in a variety of ways*, and the truly effective teacher explores all possibilities and proceeds accordingly.

Can you think of methods of teaching physical education classes other than the traditional lecture-controlled practice situation? Briefly describe a few.

"The play should be spontaneous, enjoyable, fulfilling."

_____

_____

_____

_____

_____

Focus

The formality, rigidity, and negative side effects of education are described by Leonard in the following hypothetical view of sports from the year 2000. Do you agree or disagree?

The games of limitation—which include most of the sports of Civilization—faded so rapidly after the late 1980s that the last lined area disappeared from the playfield a couple of years ago. Touch football was among the last to go. In its many permutations, football was fluid and interwoven enough to remain interesting and relevant in the new age, and some Kennedy children even now play a version of it that requires no fixed boundaries and no "officials." But the aggression it sometimes encourages leaves a bad taste in the mouths of most modern children.

Baseball, by contrast, lost its relevance long ago. Played now in four major domes across the nation before small invitational audiences, baseball may be seen on some of the lesser laservision stations. Its audience, needless to say, consists mostly of men over fifty. Suffused with sweet, sleepy nostalgia, they sit near their sets late into the night transported back to another, more complex time. Baseball, indeed, characterizes much that has passed away. Its rigid rules, its fixed angles and distances, shape players to repetitive, stereotyped behaviors. Its complete reliance on officials to enforce rules and decide close plays removes the players from all moral and personal decisions, and encourages them, in fact, to get away with whatever they can. Its preoccupation with statistics reveals its view of human worth: players are valued for how many percentage points, hits, home runs, runs batted in and the like they can accumulate. Everything is acquisitive, comparative, competitive, limiting.

Children who have played the games of expansion are hard pressed to comprehend baseball's great past appeal. As for these present games, many are improvised by the children themselves, then revised day by day. Refinement generally runs toward simplicity, elegance and an absolute minimum of rules. With no officials to intervene, the players themselves are repeatedly up against moral decision. (1968, pp. 168–169)

Assuming the existence of a variety of modes for stimulating the learner and many outlets for his response, the conscientious teacher wants to employ the most appropriate medium in any teaching situation. A classroom does not have to be a formalized structure housing a formalized teacher-student interactional process. A gymnasium need not be associated with lines, rigid rules, discipline, and formal exercises; nor should it be a fulfilling medium for athletically gifted youngsters at the expense of others. Educational leaders and instructional technologists are constantly suggesting environmental modifications, media possibilities, and communication techniques.

A Concern for Objectives

The educational scene is changing, not only because of the increased use of a variety of media to enhance the learning process, but also as a result of a growing concern with *clarification* of the teacher's classroom *objectives*.

What does the teacher *really* want the student at the end of a course to know? To do? To feel? These have been referred to as behavioral, instructional, or performance objectives in different but related sources. Can the teacher evaluate these objectives to see if they have been fulfilled? Students, administrators, parents, and a concerned public want a reasonably accurate idea of what is expected of students as they complete any given class or experience. More precise accountability is being requested of schools and school districts. The schools are supposed to produce evidence that they are satisfactorily changing student behaviors and that these behaviors can indeed be measured. Intended competencies are expressed in evaluative terms, with emphasis on the student's ability to demonstrate his learned behaviors, although not all students are necessarily evaluated on the basis of the same criteria. In turn, the individual teacher is accountable to his students in that he must design objectives that are (a) reasonably high, (b) clearly stated and not ambiguously, (c) measurable, and (d) being achieved.

Focus

The major proponent of procedural changes in this area is Robert F. Mager. In his book Preparing Instructional Objectives, *a forerunner of the proliferation of works on the subject, he briefly and simply describes the teacher's dilemma:*

Everybody talks about defining educational objectives, but almost nobody does anything about it. Books on education often stress objectives; "how-to" papers on programming list "defining objectives" as a first point; and training materials such as films and filmstrips often contain a description of the "objectives." But how often are educational units, whether large or small, prepared in response to the questions:

1. What is it that we must teach?
2. How will we know when we have taught it?
3. What materials and procedures will work best to teach what we wish to teach?

Not only must these questions be answered to instruct effectively, but the order in which they are answered is important.

The probable reason that objectives are usually stated poorly is that few people know how to proceed. This is not surprising, because little has been written on the preparation of objectives—very little for the schoolteacher. And with the all-important business of teaching occupying their capacities, it is easy for schoolteachers to feel that they have their objectives well "in mind," and that it is neither necessary nor possible to be more specific. (1962, p. v)

Do you think you are capable of writing an instructional objective? Try it. Choose an activity you might like to teach after you graduate from college. Write an objective which you think would be acceptable to Mager and your students.

Activity _____

Objective _____

_____

_____

_____

You may want to reexamine your objective and the way you stated it after you have read Chapter Six, which deals with instructional objectives.

Instructional Considerations

Let's return to the global educational scene once again. If a teacher or a school program is to be held accountable for its actions, with the burden of proof on the teacher rather than on the student, then the teacher must not only state his objectives but must sequence his instruction so as to reach those stated outcomes. It is no longer acceptable for a teacher to throw out instructional matter for students to grab as they can, memorize, and regurgitate at specified moments. Nor is it satisfactory for an instructor to throw out a ball to students in the gymnasium in the hope that they will acquire skill, change their attitudes, and develop fitness without his effectively mediating the learning process.

In order to organize his instructional sequence in a systematic way, the teacher must:

1. Know the activities (content) to be taught
2. Be aware of students' entry characteristics—their aptitudes, abilities, and skills
3. Know how to formulate and evaluate instructional objectives
4. Be familiar with research and theory in learning that are applicable to instructional methods in the form of guiding learning principles
5. Know various forms of instructional media
6. Be able to organize strategies systematically
7. Be aware of available human and other resources

Too demanding, you say? Perhaps. One of the leading objectives of this book is to convince you—the reader and future teacher— that these kinds of considerations not only contribute to effective teaching but are essential if teachers and students are to share any kind of meaningful experience.

We all are comfortable teaching as we have been taught; any attempt to change an entire approach to teaching is sure to meet with resistance. This is especially true if the alternative appears to require more effort and preparation on the teacher's part. Yet the problems faced by contemporary education and the continual, severely critical attacks on its quality demands that alternative procedures be attempted.

No area in education today can escape public scrutiny, and physical education has always had more than its fair share of detractors. We can continue to perpetuate existing programs and methodologies in our classes, but at what risk? Complaints can be registered for only so long before external controls and influences are felt. Truly professional educators recognize the glaring deficiencies of the program and of the classes as they have been and are being taught. It is time for action. Educators in a variety of specialties are attempting to upgrade the quality of their students' experiences, and physical educators are obliged to do so too.

Emphasis on Achievement

This book is dedicated to the proposition that, in the words of Leon Lessinger (1970), *students have a right to learn;* they should have basic competencies when they leave formal schooling. Although Lessinger is concerned primarily with reading and other cognitively oriented behaviors, we would like to extend his concept to all fundamental skills related to any experiences students have in school.

We also share Lessinger's belief that taxpayers have a right to know what educational results are being produced. Education is very expensive and is becoming more so every year. Those who subsidize education are justified in their quest for information which has heretofore been denied them. Unfortunately, one of the reasons for not making this information public is an embarrassingly simple one: educators do not know specifically what their students have actually learned.

We have available all sorts of interesting facts, such as that over 30 billion dollars in annual tax monies help to support almost 2 million teachers and almost 45 million students. We know how many students graduate annually from the different levels of schooling. Yet the schools have yet to measure learning changes and competencies satisfactorily. The results of education are of primary importance. Allocated resources and teaching responsibilities (for example, the number and types of course offerings) are secondary.

The history of education in all countries is replete with examples of a selection and weeding process. Successful students are promoted; the unsuccessful are failed and drop out. One hopes these selective practices will cease.

A newer, though not as yet dominant, view is that education should enable each student to develop his particular capacities to the utmost. The curriculum and instructional techniques must be adjusted to this new philosophy. The intent of education should not be to make each year of schooling more difficult but rather to develop the capabilities of the individual student.

Teachers and administrators should be more concerned with the unsuccessful student. The student who performs poorly on examinations is eventually labeled a failure, with resultant negative effects on his self-concept, personality, motivation, and learning level. As quickly as possible he will be eliminated from the school program, ill-equipped to face the realities of life. Physical education instructors sometimes show more interest in their talented students than in the less able. For the many young male teachers who want to go into coaching rather than teaching, it is easier and more rewarding to work with gifted students. But teachers should be concerned with the process of positive change for *all* students.

Focus

The following selection from A Handbook on Formative and Summative Evaluation of Student Learning, *by Benjamin S. Bloom, J. Thomas Hastings, and George F. Madaus, concerns this concept that education can and should produce meaningful change.*

Education for us is a process which changes the learners. Given this view we expect each program, course, and unit of education to bring about some significant change or changes in the students. Students should be different at the end of a unit from what they were before it. Students who have completed a unit of education should be different from those who have not had it. Although it is true that some of the differences in a learner between the beginning and end of secondary school are to be attributed to maturation, growth, and the influences of varied experiences, we are here concerned with the changes produced by education and in the last analysis determined by the school, curriculum, and instruction. . . .

In proposing these views of the education process, we are registering

"A newer, though not as yet dominant, view is that education should enable each student to develop his particular capacities to the utmost."

our faith (and there is much evidence to support it) that education can produce significant changes in learners. This is not to say that all learners will change in exactly the same way and to the same degree. Nor is it to say that all teachers, curricula, and schools will be equally effective in changing their students or will do so in the same way and to the same degree.

If the role of education is to produce changes in learners, then someone must decide what changes are possible and which are desirable. Every teacher-student interaction is based on some implicit conviction on the part of both the teacher and the student about the possibility and desirability of certain changes. It is not really possible to ignore these basic questions. Each teaching act and each learning act come out of answers to them. The teacher who does not want to state his educational objectives is merely avoiding an explicit verbal answer to these questions. His actions and interactions with students are implicit answers.

However, making verbal formulations of goals does not ensure that the implicit goals are congruent with the explicit. Nor does it ensure that the explicit goals will be realized. This is one of the uses of evaluation—to relate the actualities of student change to the stated formulation of charges sought. (1971, p. 8)

This book is dedicated to an approach that may achieve greater acceptance for physical education than has ever been realized. Greater demands will be made on teachers, but similar demands are being made in all areas of education. This approach is by no means universally accepted; legitimate as well as questionable objections have been raised. We hope, however, that the rationale given for implementing the procedures suggested in the remainder of the book will prove acceptable to you, and you will put these procedures into effect when you teach.

Consider the pressure on educators to acknowledge and act on the fact that students differ in their past experiences, aptitudes and other personal characteristics, expectations, and learning capabilities. Many of the contemporary approaches to rectifying the ills of education suggest ways of meeting individual needs, e.g., establishing individual goals, permitting students to work at their own speed, and giving them individual attention. Traditional group teaching, with group norms and group expectations, is under heavy fire.

The demand for a more personal and humanistic approach to education will probably be extremely difficult to implement. Classes dealing with cognitive skills can be programmed via computers or texts to individualize the learning process, but these methods are not so easily applied to physical activities. If we are truly to accept changing media in today's education, it becomes more than obvious that we must devise varied plans for individualized instruction. The teacher does not have to be the only avenue of information for the class. Instruction can proceed from many sources and in diverse ways, thereby enabling the effective teacher to play the role of a *manager of instruction* instead of serving as its sole source.

Many exciting solutions are being offered to combat the more apparent weaknesses in education. For example, why must the mastery of given content matter be restricted to a specified time period in a course? Why can't the minimum competencies expected be designated by the teacher, and students be allowed to take whatever time is needed to meet them? In this way slow and fast learners, more and less motivated students, can progress at their own rates. The goal of education should not be to flunk students, to have them experience failure, but to provide every possible opportunity for them to learn and achieve. Students should expect success in school, and they should attain it.

A growing number of educators are speaking in terms of educational programs based on performance or competency. The performances expected are specified in advance, and the students are supposed to demonstrate these behaviors. The emphasis is on output rather than on time limits or prescribed experiences. Students can

"Instruction can proceed from many sources and in diverse ways . . . enabling the effective teacher to play the role of a manager of instruction. . . ."

proceed on their own and select those experiences that will presumably help to fulfill their competency goals. In theory this approach to education is appealing and revolutionary. In reality developing and implementing performance-based programs entail many difficulties. Nevertheless, these programs promise to be as exciting as they are controversial.

The instructional-learning process is so complex that you may fear your future role as manager. But you now have an idea of current developments in education. Adapt to new demands and expectations. Don't give up or resort to standard, classical methods without trying to improve them.

Physical Education: Past, Present, and Future
What Is Physical Education?

At this stage in your academic preparation, unless your ideas about physical education are firmly entrenched, you may, like the majority of your peers, be confused about the nature and scope of the field. It is much easier to have set ideas and intentions at the start of a program or plan, but a better approach is to modify your concepts in the light of increased knowledge and experience until you can chart a course that will be acceptable to you. Many professors are still grappling with key issues facing the profession, or discipline; in fact, the question whether physical education is a profession or a discipline, or both, is extremely controversial in itself.

What does physical education mean to you? How do you define it?

Perhaps you stated that physical education is the teaching of athletic and/or recreational skills. Or coaching. Or a part of the educational process with goals similar to those in all other areas of education. Maybe you gave up in disgust.

The goals of physical education have traditionally been aligned with those of education in general; i.e., similar objectives have been sought through a "physical" medium. In the sixth edition of *Foundations of Physical Education* (1972, p. 7), Charles Bucher proposes that "physical education, an integral part of the total education process, is a field of endeavor that has as its aim the development of physically, mentally, emotionally, and socially fit citizens through the medium of physical activities that have been selected with a view to realizing these outcomes." Other writers have included more parameters to be fulfilled, such as the spiritual. Any sincere attempt to achieve these majestic aims would require a monumental effort on the part of the teacher.

Even though leaders in physical education have expressed definitions, meanings, and interpretations in various ways, despite the rubric and the semantics the similarities are readily apparent. Muska Mosston (1966) devoted his analysis of teaching styles in physical education to decisions made concerning physical (motor) responses, social interaction, emotional growth, and intellectual involvement; and it would appear that concept builders and practitioners have similar concerns for these areas. Specific objectives can be expressed in any of these four (or more) major categories. Mosston is unique in that he explains the relationship of various teaching styles to these main categories; the majority of textbooks discuss the nature of physical education in fairly routine and expected ways.

Focus

This brief excerpt from Harold Barrow's Man and His Movement: Principles of His Physical Education *(1971) presents a typically idealistic and global view of physical education.*

Physical education may be defined as education through big-muscle play activity such as sports, exercise, and dance where education's objectives may be achieved in part. By big-muscle activity is meant activities which involve the large muscles of the trunk, upper torso, and legs as opposed to muscles of the extremities. It is frequently referred to as gross motor activity. Like education, the physical education product must be determined. This product is a physically educated person. This value should be one of many values of the liberally educated person, and it has meaning only when it is related to the totality of the individual's life. However, the parameters of this aspect are somewhat more difficult to establish than other aspects and are not as clearly defined. (p. 15)

All is not harmonious within the field of physical education, however. A number of scholars have become disenchanted with the term physical education. They feel that it explains nothing, that it perpetuates a myth of mind-body dichotomy, and that it does not suggest the potential contributions of this discipline. The search for a unique body of knowledge, for some unique contribution we might make to education, is reflected in the efforts of those scholars to find a new name. Some of the substitutes proposed so far include biokinetics, anthropokinetics, kinesiology, and movement education.

Changing the name of a profession, or discipline, or both, does not automatically alter its content or its purposes, although, it can do so. Some scholars are attempting to substitute a title with greater academic respectability. Others would like us to be more selective in our aims, to meet fewer objectives but with much more success. Any proposed changes must be demonstrably superior to present conditions before they will be accepted. Nonetheless, "great debates" on the very essence and nature of physical education are fruitful means of scrutinizing the field, raising questions, justifying beliefs and practices, or changing them if necessary.

Focus

In Toward a New Curriculum in Physical Education *(1969) Marlin M. Mackenzie expresses the current conflict over terminology.*

Any attempt to evaluate physical education encounters a major difficulty because of the term *physical education*, which is both limiting and confusing. It is limiting because it implies that physical activities can be taught in a vacuum without giving consideration to thought and feeling. The concept of the totality and integrated nature of the human being negates the notion that education is a fractionated process of *mental* education on the one hand and *physical* education on the other. Education is concerned with the whole being and consists of learning modes that are based upon the interrelated cognitive, affective, and motor behaviors of man. There just cannot be a process called *physical education!*

The term is also confusing because it sometimes means a curriculum and at other times a body of knowledge. A curriculum is defined as the pedagogical strategies and procedures used by teachers to inspire students to learn. A body of knowledge is an integrated collection of facts, ideas, principles, concepts, and skills. The two entities are different; a curriculum is not a body of knowledge, nor is a body of knowledge a curriculum. To illustrate, there is the subject of mathematics education; there is the subject of physics and there is an educational program in physics. Does it follow that there is "physical" (or "mental") subject matter and a "physical" or "mental" curriculum? Thus the term physical education should be discarded so that confusion will not be perpetuated and views of the field will not be too constricted. (pp. 8–9)

Thus the answer to the question "What is physical education?" is by no means simple, and we do not pretend to know it. The purpose of this discussion is merely to place physical education in perspective and to examine trends, philosophical movements, and the general "state of the art and science."

A Look at the Past

Although it dates back centuries upon centuries, physical education as we know it today is probably less than a hundred years old. That is, the academic program and objectives that have prevailed in the twentieth century can be traced to Edward Hitchcock's formation of a department of physical education and hygiene at Amherst College in 1861.

In order to clarify the objectives usually advocated by physical educators, we have divided the development of physical education into five stages.

First, much of the earlier emphasis in physical education classes was on *health, body proportions* (anthropometric measurements), *exercises, physical training*, and *physical development*. Activities such as gymnastics were strongly advocated as a means of developing the body. At the turn of the century, physical objectives and physiological parameters were still primarily emphasized in programs of physical education. Thus concepts surrounding the training of the body to an ideal physical level and the maintenance of

". . . earlier emphasis in physical education . . . was on health, body proportions . . . *, exercises, physical training, and physical development."*

health launched physical education programs into school curricula. To what extent today is physical fitness or organic development deemed an important objective of our classes and programs?

Dudley Sargent, a very famous physical educator in his day (late 1800s and early 1900s), wrote that the purpose of muscular exercise was not only to contribute to health and beauty but also "to break up morbid mental tendencies, to dispel the gloomy shadow of despondency, and to insure serenity of spirit." Athletic activity was supposed to improve and develop character. As for physical training, as physical education was then called, its prime purpose was "the improvement of the individual man in structure and function."

Second, Thomas Wood called for the *replacement of physical training with physical education.* He wrote in 1893 that "the great thought in physical education is not the education of the physical nature, but the relation of physical training to complete education, and then the effort to make the physical contribute its full share to the life of the individual, in environment, training, and culture." Clark Hetherington called for a new physical education, "with the emphasis on education, and the understanding that it is 'physical' only in the sense that the activity of the whole organism is the educational agent and not the mind alone." He felt that physical education should encompass four basic processes: organic education (nutrition, fitness); psychomotor education (skills); character education (moral and spiritual powers); and intellectual education (understandings, insights).

A third stage also emerged in the early 1900s as *physical education acquired many of the characteristics of other aspects of education.* Rigidity, training, exercises, and formality gave way to more consideration for individual needs, interests, enjoyment, and expression. Sports, games, and play became the focal point of the programs. The total man was stressed; there was increased emphasis on programs that would contribute to mental, emotional, and social well-being. Measurement in physical education activities for the establishment of norms and for evaluation purposes became meaningful in the 1920s and 1930s with the pioneer work of David K. Brace, Frederick R. Rogers, and later, Charles H. McCloy.

During World War II and the Korean War interest was renewed in physical fitness and tests to measure and develop it. A fourth stage developed when physical educators attempted to *utilize* the *results of research in the psychology of learning,* in *education* and in related areas such as the *behavioral sciences* in order *to teach students more effectively.* The development of skilled performance became a dominant objective of the program. Minimum as well as maximum performance standards were established for a variety of activities.

In the fifth stage, means of *intellectualizing* the *learning* of activities, of *involving* the *cognitive process,* were pursued. So were ways of *developing favorable attitudes toward activity,* encouraging social interaction and desirable personal behavior, and providing for a sense of self-expression and fulfillment. In the past thirty years all these objectives, and others as well, have played major or secondary roles in the development of physical education programs.

Some Ideas of the Present

The confusion as to the future of physical education merely reflects the turmoil in education as a whole. Since educators are unable to agree on an orderly, meaningful, and acceptable philosophy, it is little wonder that physical education is also caught in the dilemma.

Comfort is found in tradition, and many educators see little wrong with past programs and objectives. Physical educators who resist tradition have attempted creative approaches, more or less satisfactorily. The need to make physical education an acceptable facet of education and the wish to contribute mightily to the education of youth are not necessarily met through only one approach. The pamphlet *This Is Physical Education,* produced by the American Association for Health, Physical Education, and Recreation (1965, p. 7), states that a good curriculum is one in which physical education experiences parallel the intellectual, social, physical, and emotional development of the child. The problem of designing an effective instructional approach and a program to meet its goals challenges the imagination.

The statement by the AAHPER that "today's physical education is the subject in which children learn to move as they move to learn" (p. 24) raises several questions. What kind of movements should they learn? To what extent? In what context? What should students be learning as they learn to move? What kinds of behaviors are supposed to be modified? How? Why? For what purpose? It is one thing for great minds in the profession to design the direction of programs and another for the teacher to face the unenviable task of implementing those programs. The transition from the drawing board to the playing field is not always easy.

Focus

In a more recent publication, Knowledge and Understanding in Physical Education *(1969), the AAHPER expands its viewpoint on physical education.*

The term physical education as used here is not limited to the traditional concept of a program of activities set up to teach children a prescribed number of games, dances, stunts and tumbling, gymnastics, and the like, with the hope that certain values such as physical fitness, cooperation, leadership, and similar general desirable characteristics and qualities would result. Rather, physical education is used to indicate not only a program but a process as well—a process in which the product (the student) is the point of focus. The term "education" then becomes the key word and the school program in physical education will include an emphasis on all aspects of the subject area that contribute toward total development, using the best teaching procedures known in order to accomplish the task. It will include the knowledge and understandings involved in performance, the development of skills in specific activities chosen to accomplish certain purposes, and an emphasis on the inherent and planned-for values related to the individual's developmental needs (physiological, psychological, and sociological).

It cannot be emphasized too strongly that physical education is basically an activity program, for herein lies its strength as a school subject and a teaching tool. No one can dispute the tremendous importance of physical activity in early childhood as the source of knowledge of the world around us. All of the early concepts of how to deal with space, time, and force are acquired through actually coping with these factors as the child moves in his crib, as he reaches for objects, pulls himself up, rolls over, kicks, and finally walks, runs, and climbs. As the child develops, however, the reason "why" and the recognition of "how" become increasingly important to a complete understanding of fundamental facts, principles, and procedures that assist in the development of skills of motor performance. Values related to both personal needs and the needs of others are gradually recognized by the child and play an important part in growth toward maturity. (p. vii)

Movement is the medium in today's physical education. The potential impact of the program on the student can obviously be sensed by the future physical educator. Are all the possible stated and implied objectives actually being fulfilled through the medium of movement? Let us examine the objectives proposed by the Physical Education Division Committee of the AAHPER, which was formed to write a position paper for secondary school physical education. Presented in *JOHPER* (April, 1971), these objectives will enable you to weigh possibilities, especially as you progress through this book and attempt to determine, specify, and evaluate objectives in the physical education classes that you will ultimately teach. Physical education, according to this committee, encourages the student to:

1. Develop the skills of movement, the knowledge of how and why one moves, and the ways in which movement may be organized.
2. Learn to move skillfully and effectively through exercise, games, sports, dance, and aquatics.
3. Enrich his understanding of the concepts of space, time, and force as they are related to movement.
4. Express culturally approved patterns of personal behavior and interpersonal relationships in and through games, sports, and dance.
5. Condition the heart, lungs, muscles, and other organic systems of the body to meet daily and emergency demands.
6. Acquire an appreciation of and a respect for good physical condition (fitness), a functional posture, and a sense of personal well-being.
7. Develop an interest and a desire to participate in lifetime recreational sports.

It is of interest to note those objectives of physical education that have remained essentially the same for many years, those that are relatively new, and those that are to receive top priority. But in a hundred years' time the aim of physical education has been transformed from the development of a trained and healthy body to the dynamic, integrated development of mind and body through movement.

A Look at the Future

Until recently one of the major problems in physical education has been that we have attempted to prescribe one program (the one of our choice) for all students, as if all would benefit in the same way from the same experiences and with similar objectives. As a first step in rectifying this situation, educators should realize the necessity of analyzing the program in which they are involved; determining suitable, satisfactory objectives; and suggesting alternative routes by which students may reach those objectives. Programs and objectives may vary from one class to another.

What do we really want our students to attain from their experience in physical education? Can appropriate instructional media be used to attain these objectives? Can outcomes be evaluated so that we know that objectives are in fact being fulfilled? These questions are raised to force educators to carefully think through their plans and the experiences they intend to offer to students.

Furthermore, certain procedural processes may be more desirable than others. The trend in education at present is toward self-directed and self-paced learning, individualized instruction, and independent study programs. If physical education supports such approaches to and in learning, then instructional processes must be arranged to fulfill designated class and program objectives. Thus we have identified specified, attainable, and measurable objectives; objectives to fit situations; alternative programs and learning strategies; and individualized (paced, directed, independent) instruction as immediate considerations for the future development of curricula.

Notice that no particular program is specified. No objectives are isolated and preferred. Learning strategies are not prescribed, although in general they should accommodate differences among students with respect to (a) entry skills and other characteristics, (b) abilities, aptitudes, and learning rates, (c) motivation and aspiration levels, and (d) responsiveness to various instructional environments. In other words, individualized approaches are advocated wherever possible for the realization of personal goals.

But there is structure in this apparently nonstructured approach. Despite the plea for more individual, self-realizing student experiences, the teacher is encouraged to proceed systematically, to formulate and sequence instruction in a manner designed to fulfill student objectives that have been clearly specified at the onset of the program, and to develop evaluative techniques. The appropriateness of this direction for physical education, or education in general for that matter, can be easily accepted on certain terms and rejected on others. However, it is supported throughout this book.

Let us examine another possibility for the future of physical education. George Angell, president of the State University College of Arts and Science at Plattsburgh, New York, presented a humanistic point of view in a *JOHPER* (June, 1969) article. He feels that the more important outcomes of physical education programs should be "personal identity, human worth, beauty, sensitivity, the joy of living, the art of communication," which, of course, defy measurement. He thinks we are wasting our time appraising strength, endurance, performance variables, and rules.

> Physical education is not a fact or a set of facts. It is a response—a series of responses—that initiates motion, that frees men from inertia, that expresses life in unlimited variations. Physical education is self education, the discovery of self through bodily movement, the discovery of one's unlimited capacity to express love, beauty, freedom, discipline, meaning—anything that seems beyond verbal explication. (p. 26)

Angell favors a program that encourages children to create their own activities by arousing their interest, encouraging them to participate, and stimulating questions concerning health and activity. These goals can certainly be transformed into workable instructional strategies by means of the systematic program advocated by the authors. Some of the other outcomes expressed by Angell are far less tangible, difficult to plan instruction for, and surely just about impossible to evaluate. Nevertheless, his point of view deserves consideration, for it has much merit.

We believe that students and the public are requesting clearly specified learning objectives for all students, sound and varied instructional strategies to achieve them, and assessment techniques to determine the degree to which they are achieved. A glance at the September, 1971 issue of the *JOHPER* indicates that many such programs are being formulated and are operative. The titles of these programs reveal their nature: "Independent Study Option," "Student Choice of Independent Study Units," "Self-Directed Learning," "Goal-Centered Individualized Learning," "An Elective Curriculum: Day-to-Day Choices," "The Quinmester: Extended School Year Plan," "Student-Designed Elective Course," "Contingency Contracting," "Individualized Approach to Learning," "Performance Objectives," and "Revitalizing a County-Wide Program." These experimental ripples may become waves. Programs of the future will emphasize flexible, student-oriented scheduling, individual choices, independent efforts, and self-paced decisions.

How do we plan for and execute programs departing so radically from tradition? We need a blueprint or plan. We need ideas and knowledge. We need flexibility. And we need a desire to make it all work.

Summary

At this point you should be able to identify and briefly discuss some of the major issues facing all educators today and offer possible solutions to them. Since students have different learning styles and approaches to learning situations, the schools should provide a variety of instructional media. Teachers should specify instructional objectives so that students know what is expected of them. Poor instructional outcomes can be eliminated when objectives, student abilities, learning principles, and the like are taken into consideration. And the philosophy that the school is a weeding-out place for those students who fail to reach preestablished standards should be reevaluated. Perhaps it is the teacher, and not just the student, who should be held accountable for instructional outcomes. The goal must be achievement, not failure. Physical education is faced with concerns similar to those of education as a whole. You should be able to recognize these common problems.

Having acquired some background material on historical developments, present beliefs and practices, and ideas for the future, you should be able to express some general concepts of the ideal future development of physical education programs. Physical education was originally associated with physical development and training. At the turn of the century, physical education attempted to assume many of the characteristics of education in general. The direction was toward more enjoyable activities, individual expressions of behavior, and less rigidity and formalized teaching in the schools. Although a strong physical fitness movement developed in the 1950s and early 1960s, teaching methods began to reflect research in the behavioral sciences and education. The present approach in physical education emphasizes respect for individual differences, individualized instruction, the encouragement of problem-solving experiences, and the development of the student in a variety of ways. Instructional objectives generally fall into four categories: physical development and acquisition of skill, social adjustment, attitude formation, and knowledge. In the next chapter we will attempt to categorize teaching models that will best fulfill these objectives.

Resources

Books

Education

Biehler, Robert F. *Psychology Applied to Teaching*. Boston: Houghton Mifflin, 1971. Although for teachers, the material is valuable and the approach interesting for physical educators.

Bishop, Lloyd K. *Individualized Educational Systems*. New York: Harper & Row, 1971. Practical approach for individualizing instruction, with examples drawn from innovative elementary and secondary school programs.

DeCecco, John P. *The Psychology of Learning and Instruction.* Englewood Cliffs, N.J.: Prentice-Hall, 1968. Presents a model for teaching fairly compatible with the ideas in our book. It is organized and written well, and any future teacher in any area of education can derive much value from it.

Full, Harold. *Controversy in American Education.* New York: Macmillan, 1967. Anthology of readings characterizing the major areas of controversy in American education today.

Gerlach, Vernon S., and Ely, Donald P. *Teaching and Media.* Englewood Cliffs, N.J.: Prentice-Hall, 1971. Although primarily developed to show the role of media in instruction, the contents are organized in a systems model. The approach is somewhat related to ours.

Leonard, George B. *Education and Ecstasy.* New York: Delacorte, 1968. Wants to break down the barriers of the traditional approaches to and concepts of education; heavy emphasis on making learning environments and procedures creative and various.

Lessinger, Leon. *Every Kid A Winner: Accountability in Education.* New York: Simon & Schuster, 1970. Confronts controversial issues and suggests accountability as the answer. General theme is that students have a right to learn and schools are obligated to measure proven learning.

Physical Education

Barrow, Harold. *Man and His Movement.* Philadelphia: Lea & Febiger, 1971.

Brown, Camille, and Cassidy, Rosalind. *Theory in Physical Education.* Philadelphia: Lea & Febiger, 1963. Presents many of the popular ideas of physical education; good overview of the field.

Bucher, Charles A. *Foundations of Physical Education.* St. Louis: Mosby, 1972. Monumental effort to present the nature and scope of physical education from a traditional point of view.

Daughtry, Grayson, and Woods, John. *Physical Education Programs.* Philadelphia: Saunders, 1971. Covers a wide assortment of matter related to programs and curricula.

Kroll, Walter P. *Perspectives in Physical Education.* New York: Academic Press, 1971. Although geared for graduate students and issues other than teaching, a few chapters (e.g. 1, 2, 4, 12) are appropriate to the level of our book.

Larson, Leonard. *Curriculum Standards and Foundations.* Englewood Cliffs, N.J.: Prentice-Hall, 1970. Helps place today's physical education curriculum in general perspective.

Mackenzie, Marlin M. *Toward a New Curriculum in Physical Education.* New York: McGraw-Hill, 1969. Shows dissatisfaction with physical education today and suggests ways of reconstituting it. Good for student reactions.

Mosston, Muska. *Teaching Physical Education.* Columbus, Ohio: Merrill, 1966. Overview of teaching styles with directions on how to encourage problem solving and creativity in activity settings. A must.

Oberteuffer, Delbert, and Ulrich, Celeste. *Physical Education.* New York: Harper & Row, 1962. Examines physical education in the framework of education. A classic in its time.

Pamphlets

Physical Education

This Is Physical Education. Washington, D.C.: AAHPER, 1965. A position platform on physical education from a selected committee.

Knowledge and Understanding in Physical Education. Washington, D.C.: AAHPER, 1969. A pioneer committee attempt to set forth the body of knowledge for physical education.

Articles

Education

Burns, Richard. Methods for Individualizing Instruction. *Educational Technology*, 11:55–56, 1971. Typical practical article representing concepts expressed in our text.

Physical Education

AAHPER. "The New Physical Education." *Journal of Health, Physical Education, and Recreation*, 42:24–39, 1971. Series of brief reports on innovative programs currently operating in secondary schools.

AAHPER. "Toward Program Excellence: The Physical Education Position Papers." *Journal of Health, Physical Education, and Recreation*, 42:41–53, 1971. Contains statements of objectives for different school levels, as expressed by AAHPER appointed committees.

Angell, George. "Physical Education and the New Breed." *Journal of Health, Physical Education, and Recreation*, 40:25–28, 1969. Different point of view. Good material for students to react to.

Klesius, Stephen. "Physical Education in the Seventies: Where Do You Stand?" *Journal of Health, Physical Education, and Recreation*, 42:46–47, 1971. Brief but meaningful thoughts on some critical issues in physical education.

2

Teaching Models and Theories

Anyone who is involved in education knows that many teaching approaches exist, that some are more appealing than others, and that no particular one will satisfy all students. Learning theory provides some basis for instructional processes, and teaching models more specifically suggest guidelines for understanding goal orientation and implementing teaching strategies. This chapter describes four classifications of teaching models. Since each serves a different purpose, none is considered superior to the others; the relative merits of each can only be judged in terms of the established instructional goals. Also, a teacher's priorities and preferred teaching style will determine his choice of model.

Chapter Contents

A Spectrum of Styles
The Role of Models and Theories
Four Models
Information Processing
Social Interaction
Personal Sources
Behavior Modification
Beyond Theory
Summary

Student Objectives

The student who understands the material in the chapter should, with the book closed, be able to:

1. Identify three criteria by which teaching can be analyzed
2. Discuss the potential and actual role of learning theory in the instructional process
3. Describe four families of teaching models and their relative merits
4. Give one model example of each of the following approaches in teaching physical education:
    (a) the social interaction approach

(b) the information-processing approach
(c) the personal sources approach
(d) the behavior modification approach
5. Indicate why it is important to think systematically in determining class procedures and instructional techniques

A Spectrum of Styles

From the title of this chapter, you can logically deduce that there are various approaches to the teaching of physical education. Variety can be the spice of life or an indicant of mass confusion. The rationale behind the present state of affairs in physical education should prove interesting and meaningful.

You have probably been taught physical activities in elementary school, junior high school, high school, and college. It is to be hoped that the teaching content and methods were appropriate to your age. But the philosophical and behavioral approaches of your instructors may have differed at any level or within any level of schooling. Which experiences did you enjoy the most? Which the least? Why not share your thoughts with your classmates and see if you agree as to the best and worst approaches in teaching.

There will undoubtedly be differences of opinion. Some students enjoy certain experiences and are more completely fulfilled in some circumstances than in others; some teachers advocate certain class procedures while others do not. A teacher's methodology results from the interaction of his personality, his past experiences, present knowledges and conditions, and a host of other variables. Muska Mosston has perhaps done more than anyone else in recent years to call attention to the various concepts of teaching in physical education. In his book, *Teaching Physical Education* (1966), he has identified what he calls a "spectrum of styles." A particular style may lead to the greater realization of certain class objectives than another style.

Although our purpose in this chapter is to review both traditional and contemporary teaching methods, we will not attempt to evaluate them. Each in itself or in combination with others can make a contribution, for better or worse, to the instructional situation. In the succeeding chapters, however, we will present a systematic approach which may permit, encourage, or support the inclusion of all or part of the various teaching models which represent the styles Mosston describes.

One can see that there is no one conception of teaching. Psychologists disagree as to how behavior can best be modified. Educators continually fight verbally among themselves. Philosophers and scientists have their ideas as do parents and other interested parties.

Those teaching styles that have become most popular reflect leading movements of their times. But the most important issue is not how something is taught but whether a sound basis exists for the procedures utilized.

Physical education activities can be taught in a formal setting, with the teacher playing the major role. Skills can be learned by means of drills or in trial-and-error fashion. The learning process can be programmed systematically so that learning is individualized rather than group-oriented. Student behaviors can be molded. The individual processes of exploration, problem solving, and creativity can be developed. The learning environment can encourage the mastery of skills, the attainment of information, and the development of problem-solving processes, interpersonal relations, or whatever.

In a chapter entitled "Theories of Teaching" (1964), N. L. Gage states that teaching can be analyzed according to at least four criteria: (a) teacher activities, (b) educational objectives, (c) components corresponding to those of learning, and (d) families of learning theory. *Teacher activities* include all those in which he engages; everything he does. *Educational objectives* determine the particular direction his teaching will take. For instance, major categories of objectives have been developed for the cognitive, affective, and psychomotor domains, and subsequent teaching will attempt to fulfill the designated objectives. (Obviously, the objectives for physical education lie primarily in the psychomotor domain.) *Components corresponding to those of learning* are contained in teaching; thus teaching mirrors learning. "Components of teaching might be 'motivation-producing,' 'perception-directing,' 'response-eliciting,' and 'reinforcement-providing'" (p. 276). Three *families of learning theory*—conditioning theory, identification theory, and cognitive theory—describe different kinds of learning.

We will be quite concerned later in the book with teacher activities, educational objectives, and teaching components associated with learning factors. You will be asked to analyze teaching situations in terms of student objectives expressed in precise and measurable terms. You will be provided with background material on learning conditions so that instruction will mirror research findings and theory. It will be suggested that you make a variety of preparational and actual teaching considerations in any given situation.

The Role of Models and Theories

The purpose of learning theories is to suggest logical teaching styles that will best encourage and promote learning. There is no one accepted theory. Owing to the youthfulness of psychology, theories

"Physical education activities can be taught in a formal setting, with the teacher playing the major role."

are incomplete, research findings are unresolved and contradictory, and there is no particular behavioral technology to turn to. Suggestive evidence supports various popular teaching and learning strategies. Yet there is no magic solution, no easy way out.

At first it was thought that one general learning theory could describe and explain all kinds of learning. Growing dissatisfaction with this premise led to the formulation of learning models, miniature theories applied to certain kinds of learning. In fact, one of the major criticisms of learning theory has been its lack of relevance to and impact on educational practices when they are transformed into theories of teaching. The development of the Behaviorist and Gestalt schools of thought resulted in more explicitly stated learning theories, for many of the earlier acknowledged "learning principles" were too broad and vague to be meaningful. When research data have been accumulated and synthesized, theories can be more accurately devised. Until such time the technology of teaching will try to proceed partly on the basis of common sense and logic.

If we were to examine all the possible approaches to teaching, based on common sense, teaching models, learning theories, and research, we could become either discouraged or encouraged. Perhaps one of the major problems in the past was the attempt to determine "one right way of teaching," as if such existed. A more realistic stance would be to appraise a particular approach to teaching according to the instructional goals and the kinds of students involved. It is not a question of good or bad on an absolute scale.

Consider the following situation. High school students in a physical education class are engaged in a unit entitled "basketball." Compare the following teaching approaches.

In the first case the teacher, Miss So Ciale, has brought the students together in a group. They are discussing the fact that basketball is a team sport and that it is important for the team members to cooperate in order to fulfill a common goal. Proper team attitudes, human relations, the sublimation of individual objectives for team objectives, sportsmanship, ethical conduct, and the like, constitute the substance of this interaction. Discussion and dissent are encouraged. Through carefully planned means of communication, So hopes to have the students realize that sport is an excellent vehicle for the development of cooperative behaviors, ideal human values, and good citizenship.

In our second case Mr. Cree A. Tivity is encouraging the students to discover how each—individually—can express himself on the basketball court. The prime theme is the individual experience rather than the actual organized team competition. Each student is encouraged to move and execute within the constraints of his structure, personality, and skill level. There is no intent to make the students conform to one style or do the same things at the same time. Instead, Cree feels that each student should develop his self-image, come to understand the limitations and possibilities in his movement, and experience feelings of achievement and satisfaction according to his own standards rather than those of the group.

In the third and last case Mr. Org Anization is explaining the fundamentals of basketball to his students. He believes in providing instruction that will lead to the demonstration of high levels of skill. The ability to perform well and to experience the personal pleasures of excellence in skilled movement is of top priority. Carefully planned instructions and drills follow. The students learn to execute precisely and similarly, for Org knows that excellence in basketball playing can be achieved only after many hard, dedicated, and carefully guided hours of practice.

Whose approach is best?

Why? _____

_____

_____

Obviously, all three teaching styles can be quite effective. Each can meet certain goals better than the others. If each teacher handled his assignment appropriately, we could not legitimately favor one style over another. But if particular goals were specified for a particular group of students, we would have some basis for choosing one teaching approach over the others.

And this is exactly the point taken by Bruce Joyce and Marsha Weil in their book, *Models of Teaching* (1972). After carefully reviewing and analyzing the many existing theories of teaching and learning in education as a whole, they formulated four models to represent these concepts:

1. *Information Processing*   In this model primary attention is focused on the learning process. The student's learning capacity and ability to process information, as well as his effectiveness in retrieving information accurately when required, is fundamental to the notion of information processing models. Although the usual concern is for the processing of detailed data, the accumulation of specified verbal matter or the development of well-defined skills, and the recall and application of them when required, information processing models can and have shown consideration for man as a problem-solver and even as a creator. For the purposes of this book, however, we will restrict our interpretation of information processing to traditional and fundamental roles.

2. *Social Interaction*   Here the concern is for the social development of the students. Improved social relations, the ability to relate to others, is the theme of this model. Experiences in human relations, communication, interaction, and understanding others underlie the teaching models developed in this classification.

3. *Personal Sources*   It is in this category that we will include references to problem solving and creativity. For it is here that the emphasis is on the development of self. We direct our attention to personal development, such as self image, self concept, and even the whole area of personality. The freedom and ability to express oneself motorically, to discover oneself and realize potentials and limitations, is at the heart of the movement education and problem solving approaches currently being popularized to some extent in physical education.

4. *Behavior Modification*   This final teaching model looks at man's environment and the ways his behaviors can be shaped

toward specified objectives. General learning principles affecting behavioral changes are implemented into instructional procedures. The learning variable referred to most often is that of
reinforcement, the presentation of some form of reward or acknowledgment upon the student's demonstration of behaviors
leading to the fulfillment of goals. Students are thought to generally profit in similar ways to such learning phenomena as reinforcement and other imposed modifications in the learning
environment. Behavior modification models are best represented
by the programmed instruction approach.

The general teaching model used will influence instructional
strategies and settings, and in turn the goals realized. Invariably,
a teacher formulates a priority order of goals. For example, if the
most important intended instructional outcome were the accumulation of facts and information for specific situational applications,
the information-processing model would be applied. The basic
tenets of the model would have to be understood. Classroom procedures would reveal the most efficient means of encouraging students to assimilate and master information. It is indeed possible,
and probable, that along with the preferred basic model, others
would also be implemented to some extent to help fulfill lower-
priority-order instructional goals.

Four Models

Most of the remainder of this chapter presents the four families
of general teaching models, with examples of each. As you read,
consider those methods that pertain primarily to learning theory.
Which coincides most with education objectives? Components?
Which approach is based on what the teacher does? A particular
teaching method may depend primarily on one of these variables
or any combination of them.

Information Processing

One of the major accepted functions of the school is helping students acquire skills and knowledges that will enable them to master
other tasks of a higher order. One doesn't go from ignorance to
mastery in one easy jump. A sequence of task accomplishments
leads the student to higher levels of attainment.

Instruction whose goal is the precise acquisition of skills and
knowledges is most beneficial when learning experiences reflect
an understanding of students' capacities and abilities. Such questions as how a person responds to particular stimuli, how many
stimuli or how much information he can handle effectively at any
given time, and how memory systems work are all related to
instructional decisions. It should be reiterated that straightforward

presentation and memorization of facts and specific materials, as well as specific task practice, are not the only possible interpretations of information-processing models for changing behaviors. Man must also receive input, process it, and make decisions in situations requiring problem-solving and creativity. Some excellent books have been recently published on this topic.

The Lecture Method Usually associated with information processing are more formal techniques of providing input. The lecture method is one way to provide the student with as much information as possible and to insure content coverage. It is direct and task oriented; there is little concern for those behaviors that are difficult to identify and measure. The lecture method is based on the behavioral sciences. Theoretically, its purpose is to specify with reasonable certainty (a) man's ability to distinguish among stimulus cues in a given sense modality; (b) man's channel capacity parameters—that is, how much information he can deal with at one time; (c) how much information is needed in order to learn and make correct decisions; and (d) how man retrieves learned information from his storage system. This all sounds highly technical, and some of the research and theory is. But teachers quickly learn, either intuitively or through practical experience, how and to what extent to cover material for best returns. When specific information or skills are end objectives, and especially when time is limited, teachers have tended to take more control over the student's learning experiences and to use more formal teaching approaches.

In physical education classes, lecture and drill methods suggest the teacher's goals. Instead, trial-and-error learning, a form of problem-solving behavior, is used as a sample information-processing model in this chapter. Physical education teachers usually use trial-and-error learning to help students attain specific skills, *not* to develop their problem-solving abilities.

The Drill Method The drill approach to learning reflects the influences of the Behaviorist movement in the early 1900s. Essentially, Behaviorists were concerned with manipulating the environment and predicting "normal" expected behavioral results. Little concern was given to individual differences. Stimuli could be identified, controlled, and manipulated, and corresponding responses measured. Repetition of an act led to the strengthening of "bonds." Proficiency in performance would be judged after repeated executions of the appropriate response to the given stimulus.

Many physical education teachers and coaches employ drill (Behavioristic) techniques. Classes can be organized easily. The instructor encourages all students to respond to certain cues together and in the same way—"by the numbers." Drill has its advantages and disadvantages, but it can certainly be used to elevate performance levels. Responses become habitual and a form of learning is demonstrated. Drills can be useful in the later, higher-order integration of skills if done well and varied to avoid monotony.

Focus

Illustrations of the drill techniques in physical education are easy to find. It may be the most popular style of teaching in the field, requiring attention, teacher dominance, response to commands, and predetermined levels of perfection in the execution of movements. Muska Mosston provides typical "command style" techniques, in which drill is the essence of the instructional process:

> . . . a teacher can say, "Today you will run half a mile!" or "We shall start our swimming lesson with twenty laps—slow laps!" or "When you perform the handstand today, note the place of your shoulders, make sure they are placed over the base. Also, avoid an excessive arch at the lower back!" We observe that in these statements the stimuli, the commands issued by the teacher, may take the form of either a general announcement of the selected subject matter or detailed directions for the students to follow.
>
> . . . .
>
> Whenever the process of stimulus-response functions well, identify the following features in a given lesson:
>
> 1. Organizational patterns are well executed. When the command to line up in a particular geometric form is issued, the response is practically immediate.
> 2. When attention is called for, it is there! (In the command framework of assumptions this behavior is referred to as good discipline.)
> 3. Any command for motion is followed (instantly, in most cases) by a physical response—performed either in unison or individually, primarily depending upon the nature of the activity or the traditional way in which this activity has been carried out.
> 4. A meticulous teacher will offer group or individual corrections (if the response was "wrong" in the light of his preferences). This is often done by stopping the entire class (Stimulus: "Hold it!" "Stop!" blowing a whistle, or any other agreed-upon signal. Expected response: the class stops the activity!), whereupon the teacher identifies the error, states the correction, and then gives the command to resume the activity.

5. When the lesson is over, the teacher's command stops the activity. Then follows some sort of end-lesson ceremony: a particular formation, a cheer, an announcement concerning the day's achievements, a preparatory statement about the next lesson, or the like. (1966, pp. 19–20)

Mosston does not favor this approach because he believes that it inhibits the student's social, emotional, and physical development. However, well-planned drills are undoubtedly helpful in meeting certain objectives, and lecture and drill techniques can certainly be effective in fulfilling specific task requirements. The success of any drill or lecture will depend on whether the teacher accurately evaluates his students' skills and abilities and whether he chooses his methods accordingly.

The Trial-and-Error Method Edward Thorndike, himself a behaviorist, popularized the concept of trial-and-error learning at about the same time that Pavlov, Watson, and others were investigating conditioned behavior. Extrapolating to the teaching of physical education, we might expect to see a situation such as the following: Students placed on handball courts for the first time learn to hit the ball correctly after experiencing many unexpected pitfalls. Court walls and ball bounces provide an interesting challenge, for the student gradually "stamps in" correct responses and eliminates undesirable ones from his repertoire.

Many examples have been offered of trial-and-error learning in school situations. At one end of the continuum the teacher works closely with the students, continually devising challenging situations for them to master. At the other end, the teacher unfortunately chooses to remain fairly free of responsibility. For example, after providing brief directions to a class learning volleyball, the instructor disappears into his office to take care of "more important work." The students are then left to play volleyball, picking up skills in a helter-skelter fashion. No doubt Thorndike would turn over in his grave if he could see trial-and-error learning occurring in this misdirected way. Figure 2-1 (see page 39) illustrates a problem solved by trial-and-error.

Social Interaction

A number of years ago a famous scholarly group interpreted education as the means by which we learn to live effectively within ourselves and society. If indeed we are to live fuller, richer lives, educational experiences should be geared to improving not only occupational and personal skills but social interactions and human

". . . the student gradually 'stamps in' correct responses and eliminates undesirable ones from his repertoire."

relations as well. As might be expected, social interaction teaching models suggest ways for the teacher to direct learning experiences to this end.

Once the physical educator has determined his instructional goal priorities, he can then more intelligently choose a particular model or combinations of models. Unfortunately, no social interaction model has been expressly developed for physical education, although some interesting and successful ones in other educational areas are presented by Joyce and Weil in the book referred to earlier in this chapter. One observation is pertinent here, however. Students will probably not change, develop, improve, and grow without deliberately contrived, teacher-guided activities expressly implemented to fulfill specified goals. This statement applies to any goal.

Although a "model" for social interaction per se does not exist for physical education, if the teacher thinks social objectives are important and should be realized by his students, then common sense may suggest ways by which specially designed class activities can help to produce desirable social behaviors. Extrapolation from other social interaction teaching models is not only possible but encouraged.

The noted educator John Dewey brought about many changes in educational philosophy and teaching. He believed strongly in more natural learning conditions for students, more socialized relationships among peers, problem-solving situations, and active rather than passive behaviors in the classroom. Physical education classes became more play oriented. Social outcomes and joy in experience were stressed. Classes were less structured and more student centered, with concern for individual achievements instead of group norms and the formal learning of the content of a particular discipline.

FIG. 2-1 *Trial-and-error learning*

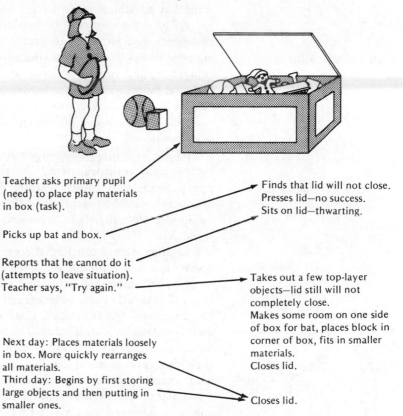

Teacher asks primary pupil (need) to place play materials in box (task).

Picks up bat and box.

Reports that he cannot do it (attempts to leave situation). Teacher says, "Try again."

Next day: Places materials loosely in box. More quickly rearranges all materials.
Third day: Begins by first storing large objects and then putting in smaller ones.

Finds that lid will not close. Presses lid—no success. Sits on lid—thwarting.

Takes out a few top-layer objects—lid still will not completely close.
Makes some room on one side of box for bat, places block in corner of box, fits in smaller materials.
Closes lid.

Closes lid.

From *The Psychology of Learning*, by Harold Bernard. Copyright © 1972 by McGraw-Hill Book Co. Used by permission of McGraw-Hill Book Co.

Personal Sources

The Behaviorism diet has satisfied many but has caused others to search for possible alternatives. Individualized, programmed instruction, in which techniques of Behaviorism are involved in the shaping of the learning process, would be one such alternative. Another would be an emphasis on movement or exploration education, in which students focus on moving their bodies within the constraints of time, space, and force with general, individualized patterns of energy. Problem solving and creativity permeate this approach as well as others.

Formalized educational experiences can make important contributions to the development of specified skills and information, and to the improvement of human relations. But what of personality development? What of self-image? What of realizing one's abilities and limitations? What of first knowing oneself?

Many sensitive observers of and participants in the athletic scene state that an athlete can discover himself through excellence in a sport. The attainment of skill, the ability to control the mind, body, and emotions in harmony are presumably one of life's most rewarding experiences. But physical education classes and athletics can be enjoyed by individuals of lesser abilities. Students can also find personal meaning in their movements. The teacher who wishes to respect individual differences and encourage the development of personal expressions of movement should use a personal sources teaching model.

Such high-order cognitive processes as problem solving and creativity, important aspects of individuality, can be developed in many physical education classes. Learning processes, rather than mastery of content, would be emphasized. Problem-solving processes could be applied to many daily experiences and used in future related situations.

Education is often considered depersonalized and nonhumanistic. Some view it as a production line, where students are thought of in terms of numbers, and specified, quantifiable skills and information are attained. These observers of the educational scene would like to change the situation by stressing individualized development, expression, and meaning. They believe that education should be an individual, not a group-centered, process. They feel that such concepts as understanding and feeling should be heavily emphasized—that is, the individual student's ability to solve a particular problem. There is not necessarily just one way to reach a solution. Realizing and respecting the nature of individual differences, can we accept and even encourage diverse approaches to solving situational problems? The teacher invents the problem, the student solves it. But there is no one right way, i.e., one level of response or expectancy to which everyone should conform. Instead each student should be encouraged to utilize his own personal resources in confronting the situation.

Personal sources models rely heavily on the development of self-knowledge. It is suggested that the more one knows about himself in a dynamic sense, the more effectively he will be able to learn and to relate to others. Individualized, personal approaches to education make great demands on the teacher, but they also help to fill a noticeable vacuum in many educational structures.

Movement Education   Elementary education is the focal point for programs in movement education. The emphasis is on the child's awareness of his body and its ability to perform gross movement patterns within certain designated situations. Presumably these experiences will lead to greater proficiency in specific athletic tasks

"The attainment of skill, the ability to control the mind, body, and emotions in harmony are presumably one of life's most rewarding experiences."

as the need arises. Problem-solving techniques are employed in this child-centered approach. The child is expected to fulfill his particular potentialities for movement; no group norms or expectations underlie the program.

Focus

A compilation of interesting thoughts on movement education by various authorities is found in Robert T. Sweeney's Selected Readings in Movement Education. *The following excerpt by Hayes Kruger indicates the nature of the underlying concepts in movement education:*

> The elements of basic movement are *space, time, force,* and *flow.* All movement has a spatial quality. It may be forward, backward, sideways, up, down, or in some combinations of directions. High, medium, and low level, as well as range of the movement, further describe its spatial relationship. We explore and thus experience the extremes of these ranges of movement, and we express them as large or small, far or near, away from or close to. The movement response to the verbalized prob-

lem presented by the teacher enables the child to contrast the differences and thus to develop wider capacity for movement possibility as the concepts of personal and general space gradually grow.

If we are able to disregard what we observe the body doing in the way of movement, we can better focus our attention either on where it is moving in relation to the available space or on what the relationship of the body parts is. No matter how the body is manipulated, it must be supported, except in brief periods of flight, such as in a leap or during a dismount from apparatus. In any case, the movement has a spatial relationship to the supporting area. Going over or under apparatus, hanging from it, or supporting oneself on it are functional ways in which movement takes place other than on the floor. And last, when an object is manipulated during a body movement it also has spatial qualities, no matter what is being done with it or to it. A ball being dribbled is at a particular level, moving within a specific range and in a particular direction, and it may range from near to far in relation to the body. Awareness of the spatial potential is important to the total learning process. Remembering that this potential is related to the perceptual awareness as well as the kinesthetic awareness of the child enables the teacher to present additional related problems, to develop discussions from observations by children of children, and to lay the foundation for analysis that becomes the basis for further exploration. Stretching, bending, twisting, turning, pushing, pulling, swinging, and many locomotor movements that transfer the body's weight to many different body parts are an outgrowth of the child's efforts to find many ways of moving. The ways of moving can be dealt with in a number of more limited problems as an outgrowth of a more general problem. As children move, they are showing the teacher what they know. If they don't move on hands and feet, perhaps they are not aware that they can or should, or perhaps they are still copying someone who always gets the "right" answers. Presenting a problem that involves movement at a low level takes care of that situation very nicely. There are many "correct" answers at this stage.

As movement-oriented teachers observing the young child at play as well as the skilled player, we can see in each case how much of the spatial potential is being realized. If we are perceptive enough, we can see many of the conceptual as well as many of the physical limitations that must be overcome in order to maximize the movement possibilities. The good coach does so as he observes his players, and he reorganizes his practice sessions accordingly. The good teacher must do so to help children develop greater awareness of their own movement potential. (1970, pp. 41–42)

Not only do a number of physical educators see value in the movement education approach. A noted educator, Charles E. Silberman, is convinced that it should be incorporated in the total scheme of revitalizing the educational process in the United States. In his provocative and controversial book, *Crisis in the Classroom*, he calls for an emphasis on "informal" physical education, which the English describe as Movement. In England in 1952–53, prescribed lessons, exercises, and a uniform program for all primary students were replaced by a program geared to develop every student's physical ability through "exploratory stages and actions which would not be the same for any two children."

Focus

Charles Silberman advocates the adoption of Movement *in this country.*
He describes the program as follows:

The most striking example of education as "the cultivation of modes of expression" is an activity that has no counterpart in American schools—something the English call "Movement," with a capital "M." One of the important aspects of the growth of informal education, the Plowden Committee writes, "has been the increasing recognition of the place of expressive movement in primary education. Children have a great capacity to respond to music, stories, and ideas, and there is a close link through movement, whether as dance or drama, with other areas of learning and experience—with speech, language, literature, and art as well as with music."

In its most fundamental sense, Movement is an attempt to educate children in the use of their bodies—to provide them with an ease, grace, and agility of bodily movement that can carry over into sports, crafts, and dance. "How many of us feel awkward, clumsy, self-conscious and embarrassed if we are called upon to perform any movement to which we are not accustomed?" John Blackie writes in *Inside the Primary School*, in a chapter interestingly entitled "Body and Soul," "How many of us have watched with envy the apparently effortless ease with which expert riders, skiers, fencers, divers, dancers, potters, woodcutters, etc., move their bodies and compared it with our own ineptitude?"

The procedure is a fascinating blend of formal and informal instruction. As a rule, an entire class participates under the teacher's direction; but precisely *how* the teacher's directions are carried out is left to each child. There is, after all, no right way or wrong way to move as if you were a snowflake, or a leaf fluttering down from a tree, which are the kinds of things children may be asked to do. The purpose, as the Plowden Committee explains it, is "to develop each child's resources as fully as possible through exploratory stages and actions which will

"Movement is an attempt to educate children in . . . an ease, grace, and agility of bodily movement that can carry over into sports, crafts, and dance."

not be the same for any two children. When these ends are pursued successfully," the Committee continues, "the children are able to bring much more to any situation than that which is specifically asked of them; the results transcend the limits of what can be prescribed or 'produced,' and lead to a greater realization of the high potential of young children."

A Movement lesson may be concerned with agility on the physical education apparatus, with skills in handling balls and other athletic apparatus, or with expressive movement of a dramatic and dancelike quality; the latter is what we are concerned with here. Barefoot and stripped to their underwear, the children assemble in the hall to learn to communicate through bodily movement—to express the whole range of feelings and emotions through the use of their hands, arms, heads, legs, torsos, and to do so with agility and ease and without self-consciousness or embarrassment.

Item: A junior school in the West Riding of Yorkshire. A class of ten- and eleven-year-old boys and girls, most of them the children of coal miners, are taking a class in Movement. The teacher, with tweed suit and British walking shoes, looking like the American stereotype of a British headmistress, calls out the directions; their execution is left to each child's imagination and ability. "Move about in a small circle, as if your body were very heavy. . . . Move about in a small circle as if your body were very light. . . . Move very quickly. . . . Move very slowly. . . . Now find a partner and make your movements in response to his, so that you are aware of what he or she is doing as well as what you are doing. . . . Speed the movements up. . . . Slow them down. . . . Make them sharp and jerky. . . . Move only your arms and body above the waist; move as if you felt very sad. . . . Move only your fingers, hands and arms, as though they were very sad. . . . Now move them as though they were very happy. . . . Find a partner and move your fingers, hands, and arms as though you were talking to each other. . . . Move about the room as though you were a butterfly. . . . Move about the room as though you were an elephant. . . . Move about in your own space as though you were a snowflake. . . . Stay in the space around you, but try to use all of it, close to the floor, above your head. . . ." All this without music, then repeated with music of various kinds. (This same school, incidentally, has the best rugby team for miles around.)

Item: A Movement class in an infant school in Bristol. A class of six- and seven-year-olds is performing a ballet of their own invention about a trip through outer space; they use only their own bodies and a few percussion instruments. One child is the earth, another the moon, several are stars. Several rocket ships go into orbit. A number of children are clouds, which at one point converge in a thunder-storm, which almost, but not quite, downs one of the rocket ships.

And so it goes; the specific approach and teaching technique vary from school district to school district and from school to school, de-pending on the tastes and talents of the teachers and heads and the influence of the local inspectors and advisors. But almost always, the children, except for an occasional shy or chubby or clumsy youngster, move with a marvelous grace and ease and apparent total lack of self-consciousness. The experience brings to mind Lillian Smith's haunting evocation of Martha Graham dancing:

> Sometimes as I have sat in the audience watching Martha Graham dance, it has seemed to me as if she were unwrapping our body image which has been tied up so long with the barbed wires of fear and guilt and ignorance, and offering it back to us: a thing of honor. Freeing, at last, our concept of Self. Saying to us, The body is not a thing of danger, it is a fine instrument that can express not only today's feeling and act, but subtle, archaic experiences, memories which words are too young in human affairs to know the meaning of.

The advocates of Movement are persuaded that the activity has a profound effect on other activities. "You don't dance to get rid of some-thing, you dance to be aware of something," Martha Graham says, and the awareness that Movement evokes seems to carry over into the children's writing, painting, and sculpture. (1970, pp. 253–255)

Problem Solving and Creativity A series of questions and prob-lems in the activity unit challenges the learner to think about alter-natives and best responses. The intellectual side of physical educa-tion is emphasized—intellectual in the sense that the learner is not told what to do but resolves the problem set before him and per-forms accordingly. Once again an individualized approach is the core, for we tend to solve similar problems in different ways. Often there is more than one correct response, although some are better than others. Whereas traditional teaching methods in physical education emphasize the skilled execution of an activity as the desirable outcome, the main feature here is the encouragement given to the learner to think about the act and to execute it in a generally acceptable way.

Focus

We return once more to Mosston's book for illustrations of the problem-solving approach. His material on soccer exemplifies the use of problem-solving techniques in mastering skills:

Let us examine this process in one of the ball games—soccer. Following the structure of subject matter, let us begin with the possible relationships between the body and the ball (excluding the use of arms, as decreed by the rules, the limiting factor of the structure).

Some problems pertaining to body-ball relationships (in motion) are:

1. What are the parts of the upper body that can be used to move the ball from point A to point B?
2. Which parts of the lower body can accomplish similar results?
3. Which parts can move the ball from point A to point B, keeping the ball rolling on the ground?
4. Which parts can move the ball from point A to point B, getting the ball slightly off the ground?
5. Which parts can move the ball, getting the ball to fly above your own height?
6. Is there another part of the body that can accomplish what you did in 4, 5, and 6?
7. Are there still other alternatives?
8. Which of the above parts of the body moved the ball the farthest?
9. Which parts can move the ball in a straight line?
10. Which can move the ball along a curved line?

Let us concentrate now on the foot:

1. Which parts of the foot can be used for moving the ball from point A to point B?
2. Can you suggest three ways of kicking and keeping the ball rolling on the ground?
3. Can you suggest two ways of kicking which will raise the ball slightly off the ground? Into higher flight?
4. Which part of the foot is best used for an accurate short kick?

5. Which part of the foot is best used for an accurate long kick?

6. Can you design three different short kicks? Three different long kicks?

7. Can you examine all previous kicks when a change of direction (of the ball) is necessary?

8. Which kicks are suitable for side kicking? Left? Right?

9. Which kicks are suitable for kicking backward?

10. Which kick is best for a "soft" kick?

11. Which kick is best for a hard kick?

12. What happens to each previous kick when it is performed after *one* preliminary step?

13. After two preliminary steps? After three?

14. After a few running steps?

15. Can you tell what will happen if you use each one of the previous kicks after a few running steps when *the ball* is in motion:
    a) Moving on the ground toward you (rolled by a partner)?
    b) Moving on the ground away from you?
    c) Moving on the ground from left to right?
    d) Moving on the ground from right to left?
    e) Review *a–d* when the ball is slightly off the ground (bounced by a partner).

16. Can you repeat each of the previous kicks using your weaker leg?

17. Which kick can you perform well with your weaker leg?

18. Are there any other dimensions that can serve as a focus for problem design?

19. Try this one—can you design problems which will focus on the speed of the ball?

. . . .

31. Can you find three ways of stopping a ball rolling toward you, using only the foot?

32. A ball rolling from the right—can you stop it using the foot in three different ways?

33. Can you find ways of stopping a flying ball using only the feet?

34. Are there any other possibilities for problems with a focus on this phase of soccer?

Similarly, other techniques of soccer are to be taught, and then [the teacher] proceeds to design problems relevant to the particular phase or technique of soccer. This has been tried for the following phases of the game: (a) heading techniques, (b) dribbling techniques, (c) techniques for use of chest and thighs in various game situations. Can you think of any other phases for which this can be tried?

A variety of tactical issues of the game can be taught by discovery. The essence of a tactical issue is *its* being a problem. In analyzing the structure of soccer, one can conceive of a great many tactical situations which call for solutions—moreover, alternative solutions to the same problems. (pp. 207–209)

Behavior Modification

For a number of years the most influential learning theory has generally involved direct situational control of students. It has been known for a long time that animal behavior can be controlled and manipulated quite easily. Humans can also be directly or subtly conditioned. The rewards and recriminations we receive for our actions influence our responses. As an approach to learning, behavior modification banks heavily on the ability of the controlled environment to shape an individual's course of action. The teacher's deliberate deployment of cues and rewards exemplifies behavior modification strategies. Armed with reinforcement weapons, he is in a powerful position to influence student behavior.

The technological revolution has led to the development of many instructional media which utilize learning principles. Such items as programmed texts and programmed machines rely heavily on such tenets as the following: (a) small chunks of material are presented to the student at one time; (b) the student works at his own speed; (c) the student knows immediately whether his response was right or wrong; (d) reinforcement is immediate; and (e) as few errors as possible are permitted to accumulate as the student progresses. These learning principles can be adopted by the teacher himself.

The behavior modification model is for the instructor who believes that students tend to proceed similarly but at different rates toward specified goals. The intent is content mastery. Not all students are expected to reach the same levels of attainment at the end of a unit or a course. But they are encouraged to proceed as quickly as they can or would like in acquiring skill or information. The learning strategies suggested in later chapters of this book might best be allied to the behavior modification approach. Let it be stated, however, that this approach is *not* the best for all instructional goals; it is elaborated upon merely as an example.

The scholar most associated with individualized programmed instruction, in which the learning process is "shaped" by immediate reinforcing techniques, is the eminent experimental psychologist B. F. Skinner. Teaching machines and programmed text materials have now been developed for many content areas; the primary concern is that each learner progress at his own speed. Subject matter proceeds from the simple to the very complex.

Difficulties arise when we try to apply these techniques to the teaching of physical education skills. It is obviously easier to program a so-called academic course than an activity course, but the latter task can be accomplished with ingenuity and effort.

Focus

Let us look at some programmed text material developed for a general course in psychology. In their book Overview of General Psychology *(Boston: Houghton Mifflin, 1966) Peter Fernald and L. Dodge Fernald include programmed text as an aid to the reader. The following example concerns the learning process:*

10-1 Learning is considered one of the most important of all the psychological processes. In almost every phase of man's diverse activities, some degree of _____ is involved.

learning

10-2 Learning plays a crucial role in making human beings psychologically different from one another and from other animals. You would not be able to read this page, recognize your friends or tie your shoe without _____.

learning

10-3 As a rule, learning is defined as some more or less permanent modification of the organism's behavior. We say "more or less permanent," because factors such as reflexes, motivation, and forgetting make it impossible to specify the time element precisely. Nevertheless, most modifications in the organism's _____ are due to learning.

behavior

10-4 A child who is jumped upon by a loudly barking dog may become afraid of dogs thereafter. If so, learning has occurred. The child has _____ to fear dogs. (1966, p. 163)

learned

Very few attempts to program movement-oriented skills have been reported in the literature. However, in material later used in a Research Quarterly *article (1968), Robert Singer and Milton Neuman describe an attempt to program instruction in tennis (each student received his own packet of instructions).*

Ball Control In order to master ball control, it is necessary to learn how the ball bounces off the racket and how to "get the feel" of the ball and the racket. Complete this section one step at a time, then proceed to the next section.

1. The forehand grip will be the grip used for all the ball-control exercises. The forehand grip is formed by shaking hands with the racket grip. The index finger and the thumb form a "V" on the top edge of the racket. The fingers are spread on the grip of the racket and the heel of the hand is at the butt. The racket is supported at the throat by the other hand to prevent undue fatigue of the racket hand. Demonstrate the forehand grip to the instructor three times. In each of the following exercises there are certain fundamental parts of the exercise which must be done correctly in order to succeed. If you find after two trials that you are not progressing well with the exercises, examine this check list of common errors, demonstrate your skill to the instructor and discuss your errors with him. Revise your pattern and continue practice.

. . . .

Forehand The forehand is a shot which is hit on the same side of the body on which the racket is held. The right-handed player hits the forehand on the right side of the body, the left-handed player on the left side.

1. The beginning stance is one in which the player has his side to the net. The right-handed player will have his left side facing the net; the left-handed player will have his right side facing the net. The feet should be shoulder-width apart, the weight on the balls of the feet, the knees slightly bent, the back straight, and the head looking toward the net. Assume this position three times before the instructor.

2. Assume the proper stance. Bring your arm back in a motion parallel to the ground and to a point almost parallel to a line made by your toes. The arm should be straight but not locked. Now, with a short step forward, swing in the same line as the backswing at an imaginary ball. Show your movement to the instructor and correct errors. Practice this movement fifteen times to this count—"ready, step, swing." Demonstrate your stroke to the instructor three times correctly.

3. The grip which should be used for hitting the forehand is the Eastern grip. It is formed by placing the hand on the racket grip as if to shake hands with the racket. A "V" is formed between the index finger and the thumb which runs up the edge of the racket when the racket face is held perpendicular to the ground. The fingers should be well spread on the grip with the heel of the hand at the butt of the racket. Demonstrate the Eastern forehand grip to the instructor three times.

4. If your stroke production is unsuccessful in the following steps, you are making mistakes in one or more of these areas. To execute the stroke correctly, the following things should be done:

   a. The wrist is kept stiff and is locked.

   b. The elbow is kept outstretched but not locked.

   c. The arm and racket are taken back in a plane parallel to the ground and at waist height.

   d. THE STROKE SHOULD RESEMBLE A ONE-ARMED BASEBALL SWING .

   e. The forward swing is in a line somewhat parallel to the ground.

   f. The racket comes around in front of the elbow. DO NOT LEAD WITH THE ELBOW.

   g. The weight transfer on the step is on the balls of the feet.

   h. The ball is contacted directly in front of the left foot (for a right-handed player) and at arm and racket length.

   i. A follow-through is made in the general flight of the ball.

   Assume the correct Eastern forehand grip and the correct beginning stance. Using the "bounce, step, swing" count, practice stroking the ball over the net three times. Ask the instructor to check all of the above points in your "bounce, step, swing" and correct any mistakes. Hit all strokes from baseline.

5. Find a partner who is at the same point in the program. Have him approved by the instructor. With your partner, practice the "bounce, step, swing" until you are able to hit fifteen of twenty balls over the net into the court. Demonstrate your proficiency to the instructor.

6. Find a partner who is at this same point in the program. Have him approved by the instructor. With your partner, practice the "bounce, step, swing" until you are able to hit fifteen or twenty balls from the right hand side of the court into the same side of the opponent's court. Demonstrate your proficiency to the instructor.

7. Find a partner who is at this same point in the program. Have him approved by the instructor. While one person *throws* the balls to the other, practice hitting exactly the same forehand stroke as that mentioned above. Each person should receive twenty balls before trading positions. When you are able to hit fifteen of twenty balls over the net and into the court, demonstrate your proficiency to the instructor. You have now completed the elementary forehand section of the programmed instruction. Proceed to the elementary backhand.

Beyond Theory

All these teaching models and others suggest some alternatives in the instructional setting. Some are better suited than others to the realization of certain objectives. Often styles are combined. All can and do produce learning. Some teachers look for alternatives in instruction, while others are more comfortable with the accepted techniques they have previously experienced. In different schools or within the same one, you may witness divers approaches to the teaching of physical education. Although in recent years the trend has been toward individualized learning, the unwieldly size of the typical physical education class encourages the teacher to assume a more dominant role in instructing for normative group behaviors by means of the "command performance."

Some teachers have attempted to reach broad-based educational and physical education objectives. Others have tried to remain true to learning theory or behavioral principles. Still others operate on a loosely structured, day-to-day basis. Many valid questions concerning student betterment have been raised by dedicated physical educators, but a satisfactory resolution still remains owing to the discrepancies in (a) stated general objectives, (b) the interpretation and use of research and learning theory, (c) teaching techniques, and (d) *the process of putting it all together.*

Regardless of the teaching method, it is usually apparent that the *way it is employed* can be open to criticism. The process of implementing a particular teaching style should reflect careful, systematic planning. It is one thing to work with objectives or theory, and something else to break down a course or unit into meaningful sequences. Are the objectives of the course specifically developed in behavioral terms? Can they be accurately evaluated to see if they are attained? Are they practical? Have limitations on space, equipment, time, and human resources been considered? If so, how? Are the students' entry skills, characteristics, and attitudes specified and taken into account when the program and objectives are formulated?

Students should know what is expected of them. Teachers are obligated to plan the students' experiences carefully. To consider all the factors impinging on the learning process and to organize experiences in a logical approach demand a good deal of effort on the part of the teacher. *Accountability*—to students, parents, administrators, and the public—requires the teacher to take his class assignments much more seriously than ever before. Where is he going, how will he get there, and how will he know when he arrives? The teacher must become a teacher–practitioner and a teacher–theorist if he is to be most effective.

One possible means for realizing teacher accountability is the *systems approach*, a methodology that forces the teacher to do more than appear before his class with enabling and terminal objectives sketched out in a traditionally hazy fashion. You will see how

"Some teachers look for alternatives in instruction, while others are more comfortable with the accepted techniques they have previously experienced."

models are developed and how they can be applied to any situation. We hope you will also conclude that the systems approach is one way to improve the public's opinion of education and educators.

Summary

We have presented a brief overview of four major educational teaching models and applied them to the teaching of physical education. The learner can benefit from any one or a combination of instructional methodologies. The issue raised is not with which class procedures are used but rather why or how. The systems approach offers concrete guidelines and a sounder operational basis for your future decisions. In the remainder of this book these guidelines will be presented in detail so that as a teacher you will be able to plan experiences that will enable your students to attain worthwhile objectives.

Resources

Books

Education

Biehler, Robert F. *Psychology Applied to Teaching.* Boston: Houghton Mifflin, 1971, Chapters 1–3. Good background material on the science and art of teaching, especially as related to learning theories.

DeCecco, John P. *The Psychology of Learning and Instruction*. Englewood Cliffs, N.J.: Prentice-Hall, 1968, Chapter 1. Serves as an excellent base for much of the material in Chapter 2 of our book.

Gage, N. L. "Theories of Teaching. " In Ernest R. Hilgard (ed.), *Theories of Learning and Instruction:* Sixty-third Yearbook of the National Society for the Study of Education. Chicago: University of Chicago Press, 1964. Excellent material on the more theoretical aspects of the analysis of teaching.

Joyce, Bruce, and Weil, Martha. *Models of Teaching*. Englewood Cliffs, N.J.: Prentice-Hall, 1972.

Silberman, Charles E. *Crisis in the Classroom*. New York: Random House, 1970. Insightful observations on the plight of education with recommendations for a change from formal to informal processes. See specifically pages 253–255 and 279–281 for support of the movement education approach in physical education.

Physical Education

Mosston, Muska. *Teaching Physical Education*. Columbus, Ohio: Merrill, 1966. Easy-to-follow analysis of teaching styles; heavy emphasis on the value of the problem-solving approach and experiences in creativity.

Sweeney, Robert T. *Selected Readings in Movement Education*. Reading, Mass.: Addison-Wesley, 1970. Background readings in movement education.

Articles   *Physical Education*

Singer, Robert, and Neuman, Milton. "A Comparison of Traditional Versus Programmed Methods of Learning Tennis." *Research Quarterly*, 39: 1044–1048, 1968.

3

An Overview of
the Systems Approach

In this chapter we will describe the development of the systems approach to teaching. This approach is quite unlike the methods previously mentioned, such as inquiry or movement, because it focuses on a much wider range of teaching activities, which should be considered part of the total instructional program.

Your understanding of this chapter is important because it sets the tone for the chapters which follow. At first you may be unfamiliar with much of the terminology, and many of the ideas may be quite different from those you have experienced in your own education. The systems approach incorporates these ideas into a consistent model which emphasizes the achievement of the individual student.

Chapter Contents

Historical Background of the Systems Approach
The Systems Approach Model
Instructional Goals
Instructional Analysis
Entry Skills, Knowledge, and Characteristics
Performance Objectives
Criterion-Referenced Evaluation
Instructional Strategies
Media Selection
Instructional Materials
Formative Evaluation
Summary

Student Objectives

The student who understands the material in this chapter should, with the book closed, be able to:

1. List at least two factors which contributed to the development of the systems approach to teaching
2. Identify the various components of the systems approach to teaching, given a diagram of the model

3. Identify the major activity associated with each component of the systems approach model
4. Describe the role of the master teacher of the future in terms of his use of the systems approach

Historical Background of the Systems Approach

In the previous chapter we referred to B. F. Skinner. Dr. Skinner has worked for several decades to determine the underlying principles by which animals and humans learn. He has found that in order to train a rat to perform a very complex task, he must first break down that task into its smallest components, and then reward the rat each time it performs correctly.

In his best-known experiment Skinner trained rats to press a bar placed in their cage. He would allow a rat to move about freely in its cage until it accidentally pressed the bar. Whenever it did so, it received a food pellet. Skinner soon found that by manipulating the rat's food, he could also manipulate its bar-pressing behavior.

In 1953, so the story goes, Dr. Skinner visited his daughter's elementary school. He observed that she and a number of the other students were having difficulty with mathematics. It occurred to him that the principles he had used to teach rats very complex tasks could also be applied to the teaching of mathematics: the content could be broken down into small units and the students could be rewarded for successfully answering increasingly complex questions. This was the beginning of the programmed instruction movement.

At one time or another you may have used either a teaching machine or a programmed instruction text, inspired by Skinner, in which you would read a sentence or two and be asked to answer a question. After writing in your response, you would either turn a crank on the machine, turn the page of your text, or move a mask down the page to learn the correct answer. This answer served as your reinforcement if you were correct or indicated your error if you were wrong. You would then proceed to the next unit of information and another question.

From this beginning a number of developments took place. Several different programming styles were soon developed, and various psychological principles were quoted to justify changes in Skinner's original format. In the early 1960s a very significant development took place when Dr. Robert Mager published his now famous book, *Preparing Instructional Objectives*. Mager's thesis was that it was not sufficient simply to analyze the desired behaviors of students and to give them small reinforcements, as Skinner had indicated. In Mager's view, the teacher should also know before he began teaching exactly what he wanted the student to be able to do.

Mager was reacting against global descriptions of "things kids should know," which could not be verified or observed in any way.

Mager has argued very persuasively that teachers should be able to describe the behaviors which they want students to exhibit as a result of having received some instruction or having learned something. Not only should the teacher be able to describe this behavior, but he should also be able to describe the conditions under which the student would be expected to do this and how well he would be expected to do it. One of the major impacts of Mager's emphasis on performance objectives has been the highlighting of the need to relate a student's progress to his teacher's actual objectives for his learning.

It became all too obvious that teachers had often indicated certain objectives for students, and then evaluated the students in a way that was completely unrelated to those objectives. Therefore, by the mid 1960s the teaching world was becoming aware of the need to specify instructional objectives, to relate evaluation to these objectives, and then to develop instructional techniques which would permit the students to achieve them.

During the mid 1960s, as a result of changes in approaches to learning and teaching, and in attitudes toward general social issues, a renewed interest developed in the area of individualized instruction. While educators had talked about the need for individualized programs since the 1920s and 1930s, they had done very little. The teacher lacked the technology or techniques to deal with students in any other way than in large groups. However, by the mid 1960s, with the advent of computers and other types of technology, and with the renewed concern for the real differences among students, attempts were made to discover techniques whereby each student could be developed to the fullest extent of his abilities. This was no longer simply a platitude but a real concern with implementing individualized procedures in the classroom setting.

Thus to B. F. Skinner's original theory of programmed learning were added the concepts of performance objectives and individualized instruction. Several educational psychologists became concerned with the development of the total concept of individualized instructional systems. The term systems is important here because it implies that the teaching-learning process is really made up of a number of activities. If the process is to succeed, all of these activities must be considered, both individually and in terms of the way in which they interact. The term systems implies that there is a predetermined relationship among the components of the instructional process; each component has its own function, and it also has an effect on other components. If a set of components form a system, one may infer that the total process is directed toward the accomplishment of a specified goal, namely, in this case, that each child will attain the learning objectives set for him. Because the psychologists were concerned with all the components in the instructional process, they referred to their models as "systems approach" models.

The Systems Approach Model

The systems approach model which will be used in this book was developed by Dick in 1968. However, the student should be aware that many such models are now being proposed by various educators. These models have many similarities; they differ primarily in the components they include or the sequence in which the components are followed. Various models are being used with a high degree of success in a variety of instructional settings at the present time. No research has been undertaken to determine the *best* model, and it is likely that such research would be unproductive because of the large number of variables which would have to be controlled.

Before moving to the next section, attempt to write your own definition or description of a "model" in the space below. (Can you think of different kinds of models? What do they have in common?)

_____

_____

_____

_____

_____

_____

Figure 3-1 shows the complete systems approach model. To be sure, the labels on it are not all self-explanatory at this point, but it should suggest to you a flow of activities which a teacher might employ in order to teach effectively. (The flow of activities is generally from left to right in the model.)

Instructional Goals

It would seem obvious that any approach to teaching should begin with some type of implicit or explicit statement about what one wants to teach. Teachers usually talk about what they are going to teach rather than what their students are going to learn. Too often they say that their instruction is in fifth grade mathematics or sophomore biology, which are really content domains and do not

FIG. 3-1 *Systems approach model for instruction*

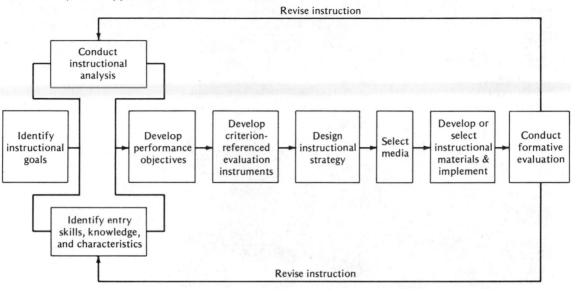

actually indicate what the student learns or is able to do as a result of having received this instruction. The statement merely says
that the teacher will somehow convey information related to these
topics.

Goal identification can be applied to any unit of instruction,
whether it is a full year in duration or merely an hour. The terminology—the way the goal is stated—can be very critical to the successful application of the systems approach model. The more specific you can be about what you want the student to be able to do,
the greater your chances of success in getting him to do it. For example, the student is more likely to learn to "successfully engage in
a game of tennis in which he appropriately observes the rules of
the game" than to be a "good tennis player." In identifying your
instructional goal, you are stating your terminal objective for the
student. It indicates what he should be able to do when he finishes
the unit of instruction.

Instructional Analysis

The systems approach model as shown in Figure 3-1 suggests that
after you have identified your instructional goal, your next step
is to conduct an instructional analysis of that goal. (Actually, the
model shows that both the instructional analysis and the analysis of
students' entry skills can be conducted at the same time.)

Have you ever wondered why certain content is included in the

"The more specific you can be about what you want the student to be able to do, the greater your chances of success in getting him to do it."

textbooks you study, or why they are presented in a particular sequence? If so, you may have realized that there is no divine or necessarily self-evident reason for including one topic and excluding another. Professional curriculum designers continually struggle with the problem of having too much material for the size of the book they wish to produce. Likewise, every teacher has favorite topics or activities which he wants to include in the instructional program. It is difficult and sometimes painful for the experienced teacher or textbook writer to subject his instructional activities to an intense analysis, because the results are sometimes dismaying.

The procedure which we are calling instructional analysis was first developed by Robert Gagné and was labeled "task analysis." Gagné has hypothesized that a student reaches an instructional goal primarily by achieving subordinate goals, and that the purpose of instruction is to move the student from one level of performance to the next higher level.

In order to perform an instructional analysis, you must begin with one of your instructional goals or terminal objectives and ask the question, "What would the learner have to do in order to perform the task after being provided with instructions?" The answer

"What would the learner have to do in order to perform the task after being provided with instructions?"

to this question would be the specification of one or more prerequisite skills or knowledges that would enable the student to learn how to do the terminal task by being given a single set of instructions.

If one were to choose a terminal objective as broad as "The learner will be able to participate effectively as a player in a nine-inning baseball game," then the answer to the question, "What would the learner have to do in order to perform the task after being provided with instructions?" might be the identification of the following five subordinate tasks or prerequisites: (a) be able to throw a baseball effectively, (b) be able to run effectively, (c) be able to bat effectively, (d) be able to catch a baseball, (e) be able to demonstrate a knowledge of the rules of baseball.

The instructional analysis does not end at this point. The same question is asked about each of the prerequisites which have been identified in the first step. For example, the instructional analysis of the subtask, "be able to bat effectively" might identify such skills as: (a) be able to stand properly in the batter's box, (b) be able to hold the bat properly prior to the release of the ball by the pitcher, (c) be able to time swing with the pitcher's motion and the path of the ball, (d) be able to manipulate the direction of the swing of the bat in order to hit the ball.

The question "What does learner have to know how to do . . ." is asked of each new skill which is identified until one reaches a very basic set of skills. In the example which we have been using, at some point in your analysis you would have reached the point at which the learner would have to be able to demonstrate some very rudimentary coordination capabilities. Figure 3-2 shows how the instructional analysis would look if it were drawn as a hierarchy of skills. Note that only one skill has been analyzed at each level. In order to complete the analysis, you would have to analyze each subtask until you identified very basic skills. Instructional analysis is not an easy process, and it is not yet in wide use in instructional situations. It should properly be considered more of an art form than an exact science, because reasonable people will analyze the same terminal objective differently. For example, you may differ with the authors in terms of the prerequisite activities associated with learning to play baseball effectively.

While teachers (and even curriculum developers) may not agree on the exact set of subordinate learning experiences which might be derived from an instructional analysis of terminal objectives, the process does have a great deal of value. For example, it clearly identifies those skills which are most critical to the achievement of the terminal objective; and by definition, those which are not identified in such an analysis would not seem to be essential. It is unlikely, for instance, that in the example we have been using one would ever identify the skill "being able to name all the teams in the

FIG. 3-2 *A partial learning hierarchy, the result of an instructional analysis*

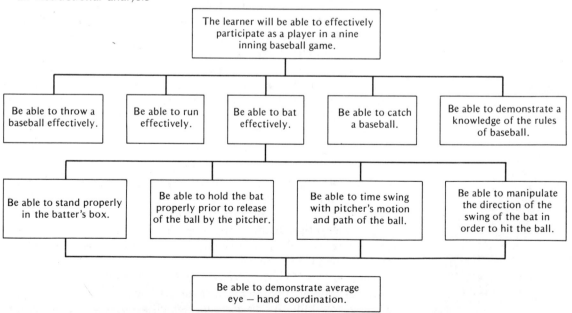

American League" as being critical in learning how to be an effective baseball player. However, without such an instructional analysis, you might well include this topic, for one reason or another, in a physical education course.

Refer again to Figure 3-2. One of the subordinate skills listed is "be able to catch a baseball." List below at least two more subordinate skills which would be identified if you asked the question, "What would the student have to know, or be able to do, so that he would be able to catch a baseball after being provided with instruction?"

_____

_____

_____

_____

_____

_____

Entry Skills, Knowledge, and Characteristics

After completing the instructional analysis, the teacher must evaluate the student's entry skills, knowledge, and characteristics. Quite often we teachers acknowledge the fact that students have different skills, abilities, and attitudes, which we refer to as individual differences, but we seldom take these differences into consideration or even attempt to measure them when we are mapping out our instruction. The easiest way out is to assume that all students know nothing about the topic to be taught and that therefore we must begin at the beginning. As a result we bore a great many students, who may have already acquired a number of the skills. Or we may err in the other direction, i.e., we may assume that students have more knowledge or skills than, in fact, they have. For example, were you tested at the beginning of this course to determine how much of the content you had already mastered? We hope so, but there is a very good chance that you were not.

Why are entering skills, knowledge, and characteristics so closely related to instructional analysis? The answer is that you can look at the instructional analysis chart and indicate very clearly the entry skills which you assume the students will have. You can draw a line across the chart at any point and then state that you assume that the students entering into this instructional activity will have already achieved those capabilities which fall *below* the line.

By using this technique, you know exactly what instructional activities to provide. You can also define the content of an evaluative test which will be administered to students entering the course and which will determine whether, in fact, they possess the skills required for entry.

Once you have assessed the students' entering capabilities, it is your obligation to decide what options would be available to the students based upon the outcome of the test. For example, will remedial instruction be available for students who do not have the assumed prerequisites for the course, or will they be asked simply not to take this instruction? You can also decide to test the students for the major skills which are to be taught in the instructional program. You must then determine whether those who have already acquired some of these skills can omit them during the instructional program.

Everything we have said to this point about evaluating entry skills and knowledge would apply equally to teachers of history, science, or physical education. Each of these teachers must be aware of the skills and knowledge a student must have in order to embark on a learning experience. However, as a physical educator you must take into account several other entry considerations, which deal primarily with students' physical and personal characteristics.

Particularly in the skills areas, the student's physical attributes will determine in large part both the point at which instruction should begin and the terminal objectives appropriate for him. Height, weight, and endurance are certainly three physical characteristics which you would not only wish to consider but would actually assess before designing an instructional program for a student in an area such as swimming or weight-lifting. Even more obvious are the special considerations for the physically handicapped.

Students' attitudes toward physical education cannot be ignored. The individual differences in this dimension are probably as great as, or greater than, those in any other dimension you may wish to consider. Any one class will contain students who are "gung ho" and students who are reluctant to participate in any activities. While the reasons for students' attitudes may be difficult to determine without a complete psychoanalysis, you must take them into consideration if the terminal objectives of your instructional program are to be achieved.

When identification of instructional goals, instructional analysis, and the assessment of entering skills, knowledge, and characteristics are regarded as the critical starting point in any teaching activity, it is clear that individualized instructional opportunities must be provided in order to meet the vastly varying needs of students.

Performance Objectives

After you have stated the instructional goals, conducted an instructional analysis, and considered the students' entry skills, you should then state their performance objectives, i.e., the specific things they will learn as a result of this instruction. Performance objectives were originally referred to as behavioral objectives, but some educators objected to the mechanistic connotation of that term, and other labels have been developed. One such label is instructional objective, which focuses on the outcome of the instruction as opposed to the process the teacher uses. We have chosen to use the term performance objective because it places the emphasis on the student and what he will be able to do when he has completed his instruction. It is fairly safe to say that you can consider all three terms—behavioral objectives, instructional objectives, and performance objectives—synonymous.

It should be stressed that the performance objectives are stated only after a thorough analysis of the instructional goals and the students' skills. Too often in the past teachers have been asked to begin by writing performance objectives. Our model suggests that writing objectives should be the fourth step in the teaching process.

As noted earlier, the real father of performance objectives is Robert Mager, who has stated that the teacher should be able to clearly describe the observable behavior he expects the student to exhibit at the end of his instruction. The performance objective should include a statement of the *conditions* under which the student will perform and the *criteria* he must achieve in order to be successful.

Let's take an example. The statement "The student should be a skillful baseball player" is certainly a reasonable instructional goal. However, it doesn't really tell us what behaviors we would be able to observe, under what conditions the student would demonstrate his skills, or what criteria we would use to judge them.

Let's take another example: "The student will be able to throw a baseball accurately." Certainly, we could observe whether or not he was throwing a baseball. But there is no description of the conditions under which he would execute this act, nor is there any real indication of the criteria for judging his accuracy.

Let's take an example of what might be considered a good performance objective: "Given a standard baseball diamond, the student will stand at home plate and throw the ball overhanded on the fly to the instructor, who is standing on first base. The instructor should be able to catch four of five throws." In this example it is clear to us under what conditions (while standing at home plate on a standard baseball diamond) the student would have to demonstrate his ability to throw. It is also clear what criterion (four of five

"The performance objective should include a statement of the conditions *under which the student will perform and the* criteria *he must achieve in order to be successful."*

throws on the fly to the instructor on first base) he will have to meet in order to demonstrate his accuracy.

Now that you understand how performance objectives are stated, you should know how they are derived. The model suggests that you look at the instructional analysis. For each of the subordinate skills that you have identified, you should write one or more performance objectives which the student should acquire. Thus you will have a number of performance objectives which students will need to master as they demonstrate their capabilities in subordinate skills.

Let's stop at this point and attempt to clarify the terminology we have been using. We have indicated that you should begin with an *instructional goal,* stated as specifically as possible. It should indicate what you would like the students to be able to do when they have completed their instruction. You then conduct an *instructional analysis* of that goal in order to determine the *subordinate skills* a student needs to learn in order to achieve the instructional goal. The result of applying instructional analysis techniques to an instructional goal is the formation of a *hierarchy* of skills to be learned by the student. After stating the required entry skills and knowledge, you should begin to write *performance objectives.* When you rewrite the instructional goal in the form of a performance objective, it then becomes the *terminal objective* for the instructional activity, i.e., the skill or knowledge toward which instructional activity is directed. The restatement of each of the subordinate skills in the hierarchy into performance objectives results in statements of *enabling objectives,* i.e., these objectives are successively achieved by the student as he works toward the terminal objective. Figure 3-3 shows the relationships among these various terms as they are applied to our baseball example.

Criterion-Referenced Evaluation

Closely related to the topic of performance objectives is that of evaluation of instruction. If you wanted to evaluate a student's 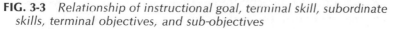 ability to throw a baseball, it would be relatively easy to define the evaluation situation. You would place the student at home plate and you would stand on first base. The student would then be asked to throw the ball to you while you are standing there. The student would be judged on his ability to throw the ball in such a way that you could catch it on the fly.

FIG. 3-3 *Relationship of instructional goal, terminal skill, subordinate skills, terminal objectives, and sub-objectives*

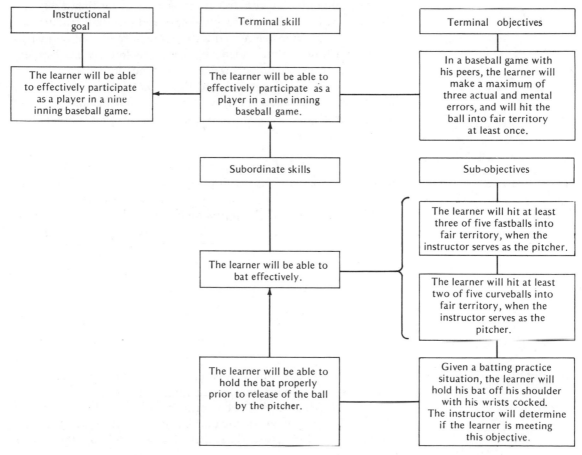

However, let's use another example of a performance objective: "Given a written test, the student will be able to answer 90 percent of the questions dealing with the rules of baseball." The objective states the conditions, the behavior, and the criteria by which the student is to be judged; what remains is the design of the actual test on the rules of baseball.

The model suggests that after stating performance objectives, you are responsible for immediately devising the instruments for evaluating the various objectives. (Note particularly that you don't first begin the instruction and then at the last minute determine how the students are to be evaluated. You must make that decision before you design the instructional activities.) An evaluation procedure which takes into account the direct relationship between the design of the evaluation instrument and the statement of the performance objective is called criterion-referenced testing. The evaluation instrument is tied directly to the criteria which you elect to use for judging the effectiveness of instruction. This concept of testing is dramatically different from the usual practice of writing a difficult test which will spread the students' scores and facilitate grading "on the curve." Criterion-referenced testing refers to evaluation instruments which determine which objectives the student has learned; it is not a means of comparing him with other students.

Though we will be returning to this topic in some detail later, it can't be stressed too much at this point that *evaluation must be tied directly to the goals and objectives for which instruction is to be provided.*

If you were told that one of the criteria for grading students in a basketball class was their ability to make six or more of ten foul shots, could you tell if the instructor was using a criterion-referenced grading system? Explain your answer below.

_____

_____

_____

_____

_____

Instructional Strategies

The model suggests that at this point you begin to engage in activities related directly to the teaching process. First you must determine the sequence of the instructional events. Have you ever wondered why you were taught one topic before being taught another? The obvious answer is that you have to know the former before you can learn the latter. This is invariably the fact when students are asked to do something more than simply memorize unrelated information.

You have already determined what must come before. This is exactly what has been done in your instructional analysis of goals. The hierarchy of goals and subgoals suggests quite clearly what the sequence of instruction should be in order to maximize the probability that the students will be able to achieve your terminal objective.

If you look at an instructional analysis, you can see that the student can proceed from a low-level skill to its related superordinate skill, and then to the next, and so on. At some point, though, several subordinate skills will come together in one superordinate skill. Therefore, you must insure that the students acquire all the subordinate skills related to any superordinate skill before providing instruction on the superordinate skill. The point here is that your sequence of instruction should be directly related to the instructional analysis you have conducted earlier.

After determining the instructional sequence, you must prepare your instructional strategy, i.e., a statement of the materials, procedures, and media you will use to help students attain each of the instructional objectives. The strategy can be a highly sophisticated analysis of the behavior associated with each performance objective, or a superficial description of the materials students will study in order to obtain each objective.

One of our sample objectives was demonstrating knowledge of the rules of baseball. Clearly, this is a cognitive skill which the student could learn in the classroom or in his own home. At one level we can describe this as a memorization, stimulus-response type of learning activity. The student simply reads the book of rules and "learns them." In fact, he is probably learning to make a large number of if-then generalizations, i.e., if a certain event happens, then this rule applies. The highest level of behavior which would be expected of the student would probably be problem solving, i.e., you could consider giving him a number of problematic situations or simulations and asking him to place himself in the role of umpire, make a decision, and describe his rationale for that decision. You will recall that with this particular objective we simply stated that the student would have to answer correctly 90 percent of the questions on a written test. Since the content of that test is unknown, the student might prepare for it in a number of ways.

Another objective was throwing a baseball. Clearly, at the beginning of the instructional session you will find great individual differences among students in psychomotor skill. You may wish

". . . you must prepare . . . a statement of the materials, procedures, and media you will use to help students attain each of the instructional objectives."

to design a practice situation in which the student receives a great deal of feedback on his performance. The instructional setting will probably be on or near a baseball field, and it will provide access to an area large enough so that the student can practice throwing the distance he will be required to throw when he is evaluated on this objective.

You should note that the type and location of the instruction are implicit in this second objective, in which the criterion by which the student is to be evaluated is clearly understood. There is a great deal more latitude in the first objective until you know exactly what the test will be like. Therefore, the model again suggests a philosophical approach to teaching and instruction different from the one you may be familiar with. The model implies that you should, in fact, be "teaching for the test," because the test is directly related to the objectives you wish to achieve. It is only when the objectives are trivial and insignificant that the teaching for these objectives becomes trivial and insignificant also.

Media Selection

After you have determined the educational objectives, designed the evaluation instruments, and planned the instructional sequence and strategy, there is one more step before you begin to develop or select the actual instructional materials: the selection of media.

"In much of your own educational experience you probably had only one medium—namely, a teacher, who served as information disseminator, question-asker, and evaluator."

In much of your own educational experience you probably had only one medium—namely, a teacher, who served as information disseminator, question-asker, and evaluator. Few media devices were available until the last several years. However, their use has increased as funds have been allocated for them and as teachers have realized that media make it possible for them to devote their time to working with individual students, while at the same time keeping the others productively engaged in the learning process.

You may feel that your use of instructional media will be limited by the resources that are available. For example, there may be available video tapes of baseball games, audio tapes on the rules of baseball, a baseball rule book, a textbook on the teaching of baseball, and some films demonstrating specific baseball skills. However, the occasion may well arise when you will want to devise instructional materials for a particular learning situation. The model suggests that you not decide immediately that the best medium to use is your own presentation. Rather, you should analyze the type of behavior expected of the student. If a performance objective clearly calls for the student to memorize information, it would make little sense for you to present that information in a lecture. It would be better to give the student a handout which he can study, reread, outline, and memorize at his own pace.

However, if the student is to model a motor skill, then his own visual representation of that skill, which can be observed over and over again, would be particularly appropriate. The student must also be provided with an opportunity to practice that skill and to receive feedback and reinforcement for his performance. Likewise, the teacher should be aware that any complex media presentation may be more confusing to the student than a simple, straightforward programmed textbook. Some studies have shown that multiple input (i.e., visual and auditory at the same time) may confuse the student. Therefore, the simplest and most direct presentation is usually preferable.

Instructional Materials

At this point you either select or develop instructional materials for use in the instructional situation. You need a great deal of experience and training in order to do this effectively. Regardless of whether these materials are developed or selected, the primary consideration is their relationship to your performance objectives and evaluation instruments.

At present, teachers design very few of their own instructional materials. However, as staff differentiation and specialization are practiced more and more in the schools, certain teachers will find that they have the time and skills required actually to develop instructional materials to meet the specific objectives of a particular instructional program.

Such materials can range from an outline for a very specific lecture to a videotape of students in a learning situation for use by preservice teachers. Clearly, it is imperative that as a developer–teacher you be continually aware of what students will have to be able to do after having received their instruction and that you try to make the instruction as clear and straightforward as possible.

Formative Evaluation

The last box in the model shown in Figure 3-1 is called formative evaluation. It is probably a new term to you, but it has become quite meaningful in the field of curriculum development. As you might assume, formative evaluation refers to the process of collecting data from students on the effectiveness of the instruction you have just provided. Notice that this does not say you should evaluate the students to see how much they have learned in order to assign them a grade. You will of course be doing this, but the primary purpose of gathering formative evaluation data is to provide you with feedback so that you can review the process which you have gone through and, if necessary, revise it.

You will notice in Figure 3-1 that two solid lines extend from the formative evaluation box all the way back to the instructional analysis and the description of entry skills, knowledge, and characteristics. This is called the feedback loop. Feedback is important for the student in order for him to know how well he is doing, and it is equally important for the teacher so that he can reexamine what he has done and make any needed improvements.

The teacher needs to evaluate the students and the procedures used in the instructional process so that he can assess the extent to which the performance objectives were achieved. He may then use the data to reevaluate the instructional analysis and the assumptions about students' entry skills and perhaps, as a result, reformulate the performance objectives and the evaluation instruments. He may change the whole instructional sequence and strategy or add other media to the instructional process.

The model suggests that the burden of providing effective instruction lies with the teacher. While he cannot learn for the student, it is mandatory that he continue to collect data and review and revise the instructional system he is operating in order to bring about maximum performance by students.

Though the model stresses the concept of formative evaluation, you should be aware that there is a second type of evaluation called summative evaluation, the comparison of two or more teaching methods to determine which one is best. This distinction between formative and summative evaluation was developed in recognition of the propensity of educators to compare old and new methods in order to make some definitive decision about which one is better. Too often innovative techniques are compared to more established ones before the former have been fully developed.

The following analogy can be made between formative and summative evaluation and the training of a racehorse. Typically, the trainer has the jockey run a horse around the track prior to a race while holding a stopwatch on him. The training and riding strategies may be changed based on the time required by the horse to run the desired distance. This is referred to as formative evaluation. The data being collected by the trainer are used to alter his training strategy for the horse, just as data collected by the teacher will be used to alter the instructional program for the students in order to make that program more effective.

To complete the analogy, when the horse appears in the starting gate with the other horses on a Saturday afternoon, the trainer begins a summative evaluation. The actual time the horse takes to run the race will not be nearly as important as his ability to beat the other horses to the finish line. He appears in the starting gate only after his training has been completed and he is at the peak of his physical readiness. The trainer has done all he can to perfect the

"The teacher needs to evaluate the students and the procedures used in the instructional process so that he can assess the extent to which the performance objectives were achieved."

jockey's riding style and to develop the best strategy for winning the race. Just as the racehorse does not compete in a race until he is fully trained, so the teacher should not attempt a summative evaluation, or a comparison between his instructional system and another, until his own has been thoroughly developed and refined to the point that it will result in the learning desired of students.

It is hoped that this analogy will help you to distinguish between formative and summative evaluation and convince you that the former is more important when you are refining and perfecting your instruction.

Summary

This chapter has presented what may have appeared to be a bewildering sequence of boxes with descriptions of assorted activities a teacher might engage in. We have anticipated that you will not feel comfortable with these concepts after a single reading; and therefore, we would encourage you to reread the chapter and reconsider the concepts and terminology which have been discussed.

Remember that what is being proposed is a model containing a number of interrelated components. The model simply states that in order to be an effective teacher, you must consider all the activities proposed in the model.

What has been presented is clearly a new role model for the teacher. It will require him to be an analyst in terms of what is being taught and somewhat of a researcher in terms of collecting data on student performance and relating it to the instructional setting. At the same time it requires him to become more of a humanist in terms of recognizing the vast array of individual differences among children and trying to provide instruction which will accommodate these differences.

One final comment may be of value. The premise of this book is that you should be knowledgeable about all the components of the instructional model which have been discussed in this chapter. However, the time may come when many of the components in the model will be developed by professional curriculum developers. For example, you may be required to follow predetermined instructional analyses of various educational goals. The state or community in which you are employed may stipulate the performance objectives and evaluation instruments you are to use. It is clear that various forms of instructional media will be available from which you will be able to select the most appropriate and that therefore, you will not necessarily need to be a curriculum development specialist. However, your overall role as a teacher will still be that of manager/decision maker. The role of the master teacher of the future will certainly be both exciting and different. This textbook has been written in anticipation of that role.

Resources

Books

Education

Glaser, R. "Toward a Behavioral Science Base for Instructional Design." In R. Glaser (ed.), *Teaching Machines and Programmed Learning II: Data and Directions*. Washington, D.C.: National Education Association, 1965.   One of the first articles to stress the systematic application of behavioral analysis to the design of instruction.

Mager, R. F. *Preparing Instructional Objectives*. Palo Alto, Calif.: Fearon, 1962.   Initiated the behavioral objectives movement.

Scriven, M. "The Methodology of Evaluation." *Perspectives of Curriculum Evaluation*. AERA Monograph Series. Chicago: Rand McNally, 1967. Scriven first identified the needed distinction between formative and summative evaluation.

Stolurow, L. M. "Social Impact of Programmed Instruction: Aptitudes and Abilities Revisited." In J. P. DeCecco (ed.), *Educational Technology*. New York: Holt, Rinehart and Winston, 1964.   Describes the renewed interest in entry behaviors that was caused by programmed instruction.

Articles

Education

Gagné, R. M. "The Acquisition of Knowledge." *Psychological Review*, 69:355–365, 1962. Early research on the use of instructional analysis.

Glaser, R. "Instructional Technology and the Measurement of Learning Outcomes." *American Psychologist*, 18:519–521, 1963. One of the first articles on criterion-referenced testing.

Skinner, B. F. "Science of Learning and the Art of Teaching." *Harvard Educational Review*, 24:86–97, 1954. Skinner's first attempt to relate his animal learning research to human learning.

Skinner, B. F. "Teaching Machines." *Science*, 128:969–977, 1958. Early article on the development of teaching machines instrumental in launching the programmed instruction movement.

4

The Systems Approach: A Case Study

In the following case study, systems approach components have been integrated into a set of procedures in order to achieve stated instructional outcomes. The outcomes were assessed, and the learners' data were used to revise the instruction so that it might be made even more effective. Though the developers in the case study did not utilize the same systems approach as that presented in this text (since there is no *one* systems approach), they did employ the general components of all systems approach models, namely, objectives, evaluation, and revision. The results dramatize the type of student achievement which can be expected when systematic procedures are applied to the instructional process.

Chapter Contents

Objectives and Prior Analyses
Strategy, Materials, and Revision
Summary

Student Objectives

The student who understands the material in this chapter should, with the book closed, be able to:

1. Describe at least three uses of student observations and performance data to develop and revise instructional materials
2. Show the effectiveness of the systems approach to teaching first aid and personal safety by describing at least three ways in which the performance of the systems approach group differed from the conventional instruction group

Several years ago the American Telephone and Telegraph Company decided to increase the efficiency and effectiveness of the first aid course which was taught to all its personnel. In order to do so, they contracted with the American Institutes for Research

(AIR) to develop a new instructional approach for the course. The AIR approach closely resembles, but is not identical to, the systems approach being used in this text. AIR used a number of creative techniques in developing an effective course which met the specifications established by AT&T.

In the description which follows, the quotations are taken from the final project report prepared by AIR (Markle, 1967).

Objectives and Prior Analyses

The general purpose or goal of the course developed by AIR was taken directly from the American National Red Cross First Aid Instructors Manual:

> The purpose of first aid training is to acquire knowledge and skill for the emergency care of the injured until a physician arrives, and to create an active interest in the prevention of accidents through elimination of the causes.

The general instructional goal for the course was that the student meet the requirement for the standard first aid certificate which is issued by the American National Red Cross, i.e., that he be able to perform basic first aid skills and to demonstrate his knowledge of how and when to apply such skills.

Rather than developing an instructional analysis of the instructional goal, the developers decided that a subject matter analysis was needed. This analysis included both the development of 500 test items based upon the American National Red Cross First Aid Instructors Manual; and, secondly, the identification of the various levels of decisions required of a person providing first aid.

Objectives summaries. The final set of approximately 500 questions was subdivided by first aid topic, such as "care for wounds," "artificial respiration," "heart attack," etc. General objectives statements were abstracted from the questions for each topic. These statements might have been produced first, had different procedures been followed. In this case, however, the statements were intended to imply only what was contained in the questions from which they were derived, and are best thought of as summaries. The questions themselves were the basis of the objectives specifications.

It would be customary to consider the objectives specification task to be complete at this point, since the conventional procedure is to specify objectives on a logical, rational basis, before engaging in instructional materials development and in empirical tryout and revision procedures. In fact, however, the set of questions and their summary statements comprised only the potential objectives from which the course objectives were to be selected. If this set of questions had been

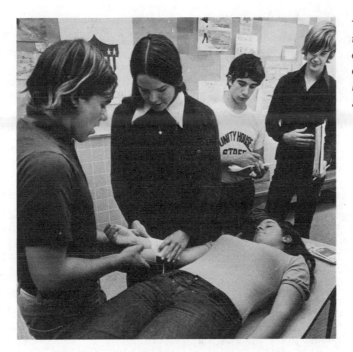

"The purpose of first aid training is to acquire knowledge and skill for the emergency care of the injured until a physician arrives. . . ."

used as the objectives, the resultant course would have been far longer and far more inclusive than existing standard first aid courses.

. . . .

Specific problems faced by this project. The limit of 7½ instructional hours made it impossible to specify exact course objectives beforehand, since the amount which can be taught is dependent to a certain extent on time available. The flexibility afforded by "self-pacing," or by homework assignments of variable length, was not available. The task was to adjust the objectives to meet the time requirements.

Several different strategies were used to accomplish this. The question, "How many of these objectives should be included?" was changed to the two questions: "Which objectives can be omitted because the students already can do those things?" and "How much of what is left can be taught in 7½ hours?" The first of these is properly part of objectives specification, and can be answered to a fair level of confidence with pretesting. The second lies in the area of course design methods. Both are empirical questions, and both are critical to making instruction as efficient as possible.

. . . .

Field testing. To answer the two questions, "What do the potential students of the new course know already?" and "What do standard first aid courses teach?", the criterion questions were administered to Bell System employees who had and had not received first aid training.

The tests were administered to approximately 800 subjects who were selected on an availability basis by the telephone company from four geographical areas. Three levels of standard first aid training were represented by the students: no training, just completed training, and training within the past five years.

Responses were tabulated on an item-by-item basis, so the error pattern of each item could be examined. Items which were seldom answered incorrectly were classed as candidates for omission from the potential objectives. Consistent error patterns were taken to indicate incorrect common knowledge which would require special attention in the new course. Typical incorrect answers included rubbing frostbite, loosening tourniquets frequently, using pulse rate as an indicator of stopped breathing, raising the feet of victims with head injuries, and removing auto accident victims from their autos immediately.

The project description indicates that although the developers did not conduct an instructional analysis, they did, in fact, analyze content skills to be demonstrated by students and used empirical techniques to determine those entry skills which almost all students would possess when they began the course. Items that almost all students could answer correctly were candidates for omission from the instruction. This is also the first of a number of instances in which decisions concerning the design of instruction were based upon empirical data rather than the best judgment of the content experts. You will see that this technique is used repeatedly throughout the project.

As a result of the testing, the project team reviewed numerous items which represented potential objectives for the first aid course and made a number of revisions. Objective comments by the students, as well as their performance, were used as a basis for clarifying these items.

Ambiguities and inefficiencies were revealed in a surprising number of the items. Identification of such problems has critical implications for the later stages of instructional materials design, because an undetected inefficient question will lead to the design of concomitantly inefficient instruction. The inefficient question will lead to teaching the student both the critical content and how to answer the unclear question, if not just the latter. When measured performance is equated with objectives, each content-related change in the measurement instrument is a detail change in the objectives. In this context, it would perhaps be better to distinguish between global, overall objectives, which may remain comparatively fixed throughout a project, and detailed, functional objectives, which are much less stable. Functional

"Typical incorrect answers included rubbing frostbite, loosening tourniquets frequently, using pulse rate as an indicator of stopped breathing. . . ."

objectives remain variable until any instructional engineering task is completed.

The paragraph above clearly highlights the interactive nature of the components within the systems approach. In essence it is being suggested that while the instructional analysis of skills may remain relatively constant throughout the project, the objectives will continue to vary based upon feedback from students in terms of their performance. Changes in test items will reflect necessary changes in objectives and vice versa.

At this point in the project, the developers had identified their instructional goal, performed their analyses and identified the entering skills which they would expect of students, specified their objectives, and developed criterion-referenced test items. The components were now in place, and the developers were ready to design the necessary instructional materials to meet their objectives.

Strategy, Materials, and Revision

The instructional strategy utilized by the project team can more appropriately be labeled a developmental strategy. The first draft of their instructional material was in the form of test items, which were presented to the students. A test question was presented to a student, and he was asked to write his answer. He was then presented with the correct answer, and he moved on to the next question. Clearly, this approximated a very difficult programmed instruction text, in which the student is given little information but is required to write in his answer and then receives the correct answer to the question.

The instructional sequence was determined by grouping the items according to the first aid topic to which they were related. Basic skill questions were presented first, then questions dealing with the use of the skill. A number of empirical tryouts of this material with students enabled the developers to identify the most difficult test items (which were now becoming the instructional materials) and those areas in which sequencing of items was necessary. While the project team was converting the test items into instructional materials, they were also developing a number of films to be used in the course.

Films and test item materials were now integrated into one package, which represented the course. A number of decisions were made based on the difficulty of the instructional material and the effectiveness of the films. It was decided that extremely difficult topics required the use of short instructional sequences or explanatory materials. These were inserted into the sequence of test and review items, and thus emphasized the instructional nature of the materials. A review unit was also added, covering those topics which had high error rates among test students as well as items relating to potential lifesaving skills. This unit was to be administered at the end of the course.

It should be noted that at this stage the course was hardly presentable. It had no introduction, no motivational material, and very little apparent continuity. All the refinement techniques applied up to this point had been concerned strictly with student performance, with no attention at all being paid to surface appearance. Certainly no conventional film producer would have shown the film to a client as an example of his skill, nor could any conventional programming criteria have been used to evaluate the workbooks which contained the brief texts, the questions, and their answers. These materials bore about the same superficial relationship to the final course as the concrete supports, still encased in their forms, bear to a completed bridge.

Course tryout, Version 1. Version 1 was administered in Atlanta, Georgia, to ten trainees who had received no prior standard first aid training. . . . Total instructional time for Version 1 was approximately twelve hours, with the films taking two hours, practice sessions taking three, and workbooks taking seven. It was obviously hard work for the trainees, who commented freely that "this is sure harder than high school." Mild complaints accompanied the distribution of the workbooks in the last few hours of the course. The films, although hardly exciting, were obviously looked forward to as respite from the workbooks, which could be described as an extreme form of "brute force" programming. The students were observed to adopt a style of working quite different from that which is normally observed with low error-rate programs. They would read a question, puzzle about its answer, make a guess at it (usually part of the complete answer), then study the correct answer given on the next page. Obviously the answers were not serving the occasional confirmation or answer-checking function they serve in conventional low-error programed instruction, but were often serving as initial instruction. That this worked, if inefficiently, is shown by the test results presented in [Table 4-1].

The no-instruction control group and the group which received Version 1 of the new course were randomly selected from the same pool of newly hired employees. Several last minute substitutions in both groups by the telephone company upset the randomness, but did not introduce any identifiable biases. . . . The control groups which had received standard first aid training were randomly sampled from the appropriate cells of the earlier administered field testing, and represent a pooling of results from numerous different standard first aid courses. . . . All testing was done immediately on completion of the course involved. The no-instruction control group was tested at the same time as the Version 1 group.

Revision of Version 1. The major goal of the revision was to reduce instructional time from twelve hours to nearer the target of 7½ hours without sacrificing the effectiveness of Version 1. Data obtained from the Atlanta tryout included observations of performance in the

TABLE 4-1 *Test form I comparisons of no instruction, standard course, and Version 1 of the new course; maximum possible score: 351 points*

	No instruction	Standard course	Version 1
Course length	0 hrs.	10 hrs.	12 hrs.
Mean score	109	132	267
Standard deviation	30	36	29
Lowest score	41	39	219
Highest score	149	244	325
Range of scores	108	205	106
Number of subjects	10	30	10

practice sessions, responses to questions in the workbooks, responses to test Form I, and student comments.

It was evident that explicit instructor guides would be required. Practice session guides were prepared for Version 2. These consisted of exact scripts for the instructor to read to the students during each practice session, and checklists on which he could check off skill points. The scripts were intended both to help the instructor organize the practice sessions, and to set up the situation so some testing of conceptual material would take place. For example, the instructor was told to say, "The victim has a serious wound on his forearm. Do what you would do first to stop the bleeding," instead of, "Now demonstrate direct pressure and elevation for a serious wound on the victim's forearm."

The workbooks were extensively revised, with strong emphasis on time-saving techniques which would not sacrifice instructional quality. Consistent correct answers on a single point within the course, in the review section, and on the final tests, were used as evidence of redundancy, and some material on these points was eliminated, as was done earlier when the separate criterion question sequences were combined. Questions which were answered uniformly correctly, but which involved content judged to be of less than critical importance, were converted to statement form in order to eliminate time-consuming response requests. This had the effect of restricting the material on which a student would repeatedly be tested on critical, high value items. Treatments of points on which consistent errors were made were of course modified to reduce the probability of these errors.

It should be noted that the project team based all their revisions on data received from students, including observation of their actual performance in the field trial, their responses to questions written in the instructional material, their responses to the criterion tests, and their comments. It can be noted in Table 4-1 that the average score for the Version 1 students on the final examination was 267 out of 351 points. This is more than double the score received by students who had already taken the standard course. Similarly, the instructional period required twelve hours, and the target time in which the course was to be administered was seven and one-half hours. Therefore, the data were used to identify ways in which performance could be increased and time saved in the instructional process. In addition, motivational materials were introduced in the course for the first time in the form of an introductory set of vignettes, which served as organizational and introductory materials for course objectives.

"The workbooks were extensively revised, with strong emphasis on time-saving techniques which would not sacrifice instructional quality."

Course tryout, Version 2. Newly produced 35mm color films were combined with the practice session guides and the revised workbooks in a manner similar to Version 1, and tested on two trial classes in San Francisco. One tryout was informally administered by the project staff; the second was formally administered by a Bell System first aid instructor. . . .

Student performance in the practice sessions was noticeably improved. Most of the procedural skills errors which had been frequent in the tryout of Version 1 were eliminated by the new film. Total instructional time was reduced from twelve hours to nine hours. This time saving resulted from the greater efficiency of the workbooks, and from the much closer interrelationship achieved between the films and the workbooks in Version 2. Minor changes in narration, which added virtually no time, had in some cases permitted us to eliminate some workbook materials, and in others enabled us to avoid adding material. Overall comparison data for Version 2 are presented in [Table 4-2] .

Revision of Version 2. Workbooks were revised according to the same strategies used previously, with further attention to time-saving changes in response requests. Interrelationships between films and workbooks were further refined. Questions on material taught in the

TABLE 4-2 *Test form I comparisons of no instruction, standard course, and Versions 1, 2, and 3 of the new course; maximum possible score: 351 points*

	No instruction	Standard course	Version 1	Version 2	Version 3
Course length	0 hrs.	10 hrs.	12 hrs.	9 hrs.	7.5 hrs.
Mean score	93	132	267	268	278
Standard deviation	32	36	29	22	16
Lowest score	27	39	219	234	251
Highest score	158	244	325	298	302
Range of scores	131	205	106	64	51
Number of subjects	30*	30	10	9	8

* The control groups reported in Tables 4-1 and 4-2 are the same, except for the addition of 20 cases to the no-instruction Form I group. These additional cases were randomly sampled from the field testing, numerically scored, and added to provide a larger N, at the request of the sponsor.

preceding films and practice sessions were placed at the beginning of each printed lesson, with the pages of instructional text and their questions following. The pages of instructional text, which had previously been grouped together, were split into smaller sections and distributed more evenly throughout the question sequences. The resulting overall pattern for each workbook lesson was (1) initial question sequence on preceding films and practice sessions, (2) brief page of text on more material, (3) question sequence on new material, (4) brief page of text on more material, (5) question sequence, etc. The format of each question sequence was as before, with a question on one page, the answer on the next. In addition to these revisions, subject matter changes suggested by the Red Cross were incorporated.

Course tryout, Version 3. The final version, as published, was tested in the Pacific Telephone and Telegraph Company. The class was administered by a Bell System instructor who had been given a fairly thorough preparation, and was completed within the prescribed seven and one-half instructional hours. Comparisons on test Form I are presented in [Table 4-2]. . . .

Limitations of the data. The data reported [in Table 4-2] provide ample evidence that the new course produced improved performance on written tests given soon after instruction. Questions remain, however, about real world performance, which cannot be answered within the limits of an initial development project. . . .

The Red Cross does have a great deal of accumulated experience with real world first aid problems, so there is good reason to believe that the course is related to lifesaving and injury-treating behavior in emergency situations. But no direct test validation data can be provided. Since the course will be given to large numbers of people, however, there exists the possibility that data on first aid incidents which involve people who have been trained by the new course can be collected. These data may eventually be used to evaluate the short term instructional goals of the course. . . .

Relative Power of the Developmental Methods. The comparative value of the methods used in this project cannot be assessed with the overall course comparison data which were obtained. A cost effectiveness analysis of the methods used would require an extensive experimental rather than developmental effort. It is nonetheless worthwhile to comment on some aspects of these methods.

Although the main innovative interest of the project lies in the wide range of empirical methods used, the importance of the initial analysis of first aid in decision-making terms must be emphasized. It is critical not that the analysis was in terms of decision making *per se,* but rather that the analysis systematically shifted emphasis from content to behavior, crosscutting existing classification categories. It appears that the discovery of an analysis scheme which crosscuts existing content-related classifications is of critical importance to projects of this sort.

Techniques for the elimination of common knowledge material were undoubtedly major contributors to the large differences in scores found between the control groups and groups taught by the new course. Standard first aid instruction contains a large amount of common knowledge material; thus, wide range tests show small differences between untrained and conventionally trained students. Because so much instruction in most subject matters is hamstrung by the reiteration of already known material, this aspect of the empirical specification of objectives deserves wide general application.

The successive approximation of instructional materials, starting from criterion questions alone, eliminated the need for initial judgments about what instruction students would require—judgments which are typically made with inadequate information. This procedure also made it possible to exercise careful control over the material which eventually was included in the course. This control is particularly important when time constraints are involved, and when several media of instruction are to be integrated. . . .

It is interesting to note that little attention was paid directly to the question of medium selection. Apart from the *a priori* decision to film the skills demonstrations, medium selection decisions were typically made for very small components of the course, in the context of specific data-based requirements and physical constraints. It appears that the highly detailed empirical work on the basic levels of objectives specification and instructional requirements determination eliminated the need for large scale decision making on the medium selection level. Design decisions made at any one level of development have surprisingly far-reaching consequences at other levels—consequences which suggest that the sequence of decision making alone is well worth further attention, apart from questions of what kinds of evidence are to be used for each decision.

Summary

While the procedures used by the AIR project staff did not follow the systems approach model which is presented in this text, they do demonstrate the efficacy of systematic analytic procedures and of using learning data from students. The project description clearly demonstrated the need for course objectives and criterion-referenced test items, and their interrelationship.

The project clearly demonstrates the value of basing course decisions on students' performance data. Three different revisions of the course have been reported here. Certainly the first version, by almost any standard, would be judged, on the basis of student performance, to be quite successful. However, as is demonstrated in Table 4-2, successive versions became even more satisfactory. In essence, the same student performance, namely, 278 of the 351 points, was maintained, while the length of the course was almost halved. It is also significant to note that the standard deviation of the scores (the extent to which student scores were spread around the average score) was also halved in comparison with the no-instruction group. The lowest score received on Version 3 was 251, while the highest score obtained by persons in the standard course was 244. All the scores of students in Version 3 were higher than the highest score obtained by the students taking the standard course.

This study is particularly apropos for students in physical education, not only in terms of the content which is often taught by the physical education teacher, but also in terms of the combination of cognitive and psychomotor skills which were learned by the first aid trainees. Thus some of the problems faced by the developers will be similar to those you will face in dealing with these same types of topics.

We realize that it is unlikely that the physical education teacher will ever have the resources available to him to participate in the type of development project which has been described here. AT&T spent several hundred thousand dollars in order to obtain instructional materials which met their desired objectives. However, this study was conducted in 1967, when many of the techniques which were utilized were being pioneered by AIR. The practicing teacher

"Design decisions made at any one level of development have surprisingly far-reaching consequences at other levels. . . ."

can now use these same techniques, admittedly on a smaller scale, to increase his effectiveness by focusing on the behaviors of students and on the empirical evidence of the effectiveness of an instructional program.

Resources

Books

Education

Banathy, B. H. *Instructional Systems.* Palo Alto, Calif.: Fearon, 1968.

Bandura, A. "Behavioral Modifications through Modeling Procedures." In L. Krasner and L. P. Cellmann (eds.), *Research in Behavior Modification.* New York: Holt, Rinehart and Winston, 1965.

Bloom, B. S., Engelhart, M. D., Furst, E. J., Hill, W. H., and Krathwohl, D. R. *Taxonomy of Educational Objectives, Handbook I: Cognitive Domain.* New York: McKay, 1956.

Briggs, L. J. *Sequencing of Instruction in Relation to Hierarchies of Competence.* Pittsburgh: American Institutes for Research, 1968.

Briggs, L. J. *Handbook of Procedures for the Design of Instruction.* Pittsburgh: American Institutes for Research, 1970.

Briggs, L. J., Campeau, P. L., Gagné, R. M., and May, M. A. *Instructional Media.* Pittsburgh: American Institutes for Research, 1967.

Churchman, C. W. *The Systems Approach.* New York: Dell, 1968.

Cronbach, L. J. *Educational Psychology* (2nd ed.). New York: Harcourt, Brace & World, 1963.

DeCecco, J. P. *The Psychology of Learning and Instruction.* Englewood Cliffs, N. J.: Prentice-Hall, 1968.

Note: In the interests of saving space, the extensive list of resources presented here is not annotated.

Elam, S., and Swanson, G. J. (eds.). *Educational Planning in the United States*. Itasca, Ill.: Peacock, 1969.

Fitts, P. "Perceptual-Motor Skill Learning." In A. W. Melton (ed.), *Categories of Human Learning*. New York: Academic Press, 1964.

Fleishman, E. A. "The Description and Prediction of Perceptual-Motor Learning." In R. Glaser (ed.), *Training Research and Education*. Pittsburgh: University of Pittsburgh Press, 1962.

Follettie, J. "Effects of Training Response Mode, Test Form, and Measure on Acquisition of Semi-Ordered Factual Materials." Research Memorandum, HumRRO Division N. 4 (Infantry), Ft. Benning, Ga., 1961.

Gagné, R. M. "Criterion Research in Curriculum Promotion of Learning." In R. Tyler, R. Gagné, and M. Scriven (eds.), *Prospecti of Curriculum Evaluation*. Chicago: Rand McNally, 1967.

Gagné, R. M. *The Conditions of Learning* (2nd ed.). New York: Holt, Rinehart and Winston, 1970.

Glaser, R. "Psychology and Instructional Technology." In R. Glaser (ed.), *Training Research and Education*. Pittsburgh: University of Pittsburgh Press, 1962.

Glaser, R., and Reynolds, J. H. "Instructional and Programmed Instruction: A Case Study." In C. M. Lindwall (ed.), *Defining Educational Objectives*. Pittsburgh: University of Pittsburgh Press, 1966.

Gronlund, N. E. *Stating Behavioral Objectives for Classroom Instruction*. New York: Macmillan, 1970.

Kibler, R. J., Barker, L. L., and Miles, D. T. *Behavioral Objectives and Instruction*. Boston: Allyn & Bacon, 1970.

Krathwohl, D. R., Bloom, B. S., and Masia, B. B. *Taxonomy of Educational Objectives, Handbook II: Affective Domain*. New York: McKay, 1964.

Lysaught, J. P. *A Guide to Programmed Instruction*. New York: Wiley, 1968.

Mager, R. F. *Preparing Instructional Objectives*. Palo Alto, Calif.: Fearon, 1962.

Mager, R. F., and Beach, K. M. *Developing Vocational Instruction*. Palo Alto, Calif.: Fearon, 1967.

Payne, D. A. *The Specification and Measurement of Learning Outcomes*. Waltham, Mass.: Blaisdell, 1968.

Slack, C. W. *Lesson Writing for Teaching Verbal Chains: Order in Which Elements Are Taught*. Stamford, Conn.: TOR Educations, 1964.

Smith, R. G. *An Annotated Bibliography on the Design of Instructional Systems*. Alexander, Va.: Human Resources Research Organization, Technical Report #66-18, 1966.

Physical Education

Karsner, M. G. "An Evaluation of Motion Picture Loops in Group Instruction in Badminton." Ph.D. dissertation, State University of Iowa, 1953.

Plese, E. R. "A Comparison of Videotape Replay with a Traditional Approach in the Teaching of Selected Gymnastic Skills." Ph.D. dissertation, Ohio State University, 1967.

Singer, R. N. *Motor Learning and Human Performance.* New York: Macmillan, 1968.

Wyness, G. B. "A Study of the Effectiveness of Motion Pictures as an Aid in Teaching a Gross Motor Skill." Ph.D. dissertation, University of Oregon, 1963.

Articles

Education

Abel, T. M. "The Influence of Social Facilitation on Motor Performance at Different Levels of Intelligence." *American Journal of Psychology*, 51:379–388, 1938.

Ausubel, D. P., and Youssef, M. "Role of Discriminability in Meaningful Parallel Learning." *Journal of Educational Psychology*, 54:331–336, 1963.

Bruner, J. S. "Needed: A Theory of Instruction." *Educational Leadership*, 20:523–532, 1963.

Canfield, A. A. "A Rationale for Performance Objectives." *Audiovisual Instruction*, 13:127–129, 1968.

Campbell, V. N. "Self-direction and Programmed Instruction for Five Different Types of Learning Objectives." *Psychology in the Schools*, 1:359–384, 1964.

Cox, J. A., and Boren, L. M. "A Study of Backward Chaining." *Journal of Educational Psychology*, 56:270–274, 1965.

Craik, M. B. "Writing Objectives for Programmed Instruction or Any Instruction." *Educational Technology*, 6:4, 1966.

Cyrs, T., and Lowenthal, R. "A Model for Curriculum Design Using a Systems Approach." *Audiovisual Instruction*, 15:16–18, 1970.

Eiss, A. F. "A Systems Approach to Developing Scientific Literacy," *Educational Technology*, 10:36–40, 1970.

Engman, B. D. "Behavioral Objectives: Key to Planning." *Science Teacher*, 35:86–87, 1968.

Gagné, R. M., and Bassler, O. C. "Study of Retention of Some Topics of Elementary Nonmetric Geometry." *Journal of Educational Psychology*, 54:123–131, 1963.

Gagné, R. M., and Paradise, N. E. "Abilities and Learning Sets in Knowledge Acquisition." *Psychological Monographs: General and Applied*, 75 (no. 518), 1961.

Gilbert, T. F. "Mathematics: The Technology of Education." *Journal of Mathematics*, 1:7–73, 1962.

Grieder, C. "Is it Possible to Word Educational Goals?" *Nations Schools*, 68:10, 1961.

Haverman, M. "Behavioral Objectives: Bandwagon or Breakthrough?" *Journal of Teacher Education*, 19:91–94, 1968.

Hoover, W. F. "Specification of Objectives." *Audiovisual Instruction*, 12:597, 1967.

Kapfer, P. G. "Behavioral Objectives in the Cognitive and Affective Domains." *Educational Technology*, 8:11–13, 1968.

Leavitt, H. B. "Dichotomy between Ends and Means in American Education." *Journal of Education*, 141:14–16, 1958.

Lehmann, H. "The Systems Approach to Education." *Audiovisual Instruction*, 13:144–148, 1968.

Lewey, A. "Empirical Validity of Major Properties of a Taxonomy of Affective Educational Objectives." *Journal of Experimental Education*, 36:70–77, 1968.

Mager, R. F. "On Sequencing of Instructional Content." *Psychological Reports*, 9:405–413, 1961.

Ojemann, R. H. "Should Educational Objectives Be Stated in Behavioral Terms?" *Elementary School Journal*, 68:223–231, 1968.

Palmer, R. R. "Evaluating School Objectives." *Education Research Bulletin*, 37:60–66, 1958.

Popham, J. W. "Tape Recorded Lectures in the College Classroom." *AV Communication Review*, 9:109–118, 1961.

Popham, J. W. "Tape Recorded Lectures in the College Classroom, II." *AV Communication Review*, 10:94–101, 1962.

Popham, J. W., and Baker, E. L. "Measuring Teachers' Attitudes Toward Behavioral Objectives." *Journal of Educational Research*, 60:453–455, 1967.

Pressey, S. L., and Kinzer, J. R. "Auto-elucidation Without Programming." *Psychology in the Schools*, 1:359–365, 1964.

Scandura, J. M. "Prior Learning, Presentation Order, and Prerequisite Practice in Problem Solving." *Journal of Experimental Education*, 34: 12–18, 1966.

Skinner, B. F. "Teaching Machines." *Science*, 128:137–158, 1958.

Stufflebeam, D. L. "Toward a Science of Education Evaluation." *Educational Technology*, 8:5–12, 1968.

Trow, W. C. "Bahavioral Objectives in Education." *Educational Technology*, 7:24, 1967.

Worthen, B. R. "Toward a Taxonomy of Evaluation Design." *Educational Technology*, 8:3–9, 1968.

Zajonc, R. B. "Social Facilitation." *Science*, 149:269–274, 1965.

Physical Education

Brown, H. S., and Messersmith, L. "An Experiment in Teaching Tumbling With and Without Motion Pictures." *Research Quarterly*, 19:304–307, 1948.

Gasson, I. "Relative Effectiveness of Teaching Beginning Badminton With and Without an Instant Replay Videotape Recorder." *Perceptual and Motor Skills*, 29:499–502, 1969.

Gray, C. A., and Brumbach, W. B. "Effect of Daylight Projection of Film Loops on Learning Badminton." *Research Quarterly*, 38:562–569, 1967.

Howell, M. L. "Use of Force-time Graphs for Performance Analysis in Facilitating Motor Learning." *Research Quarterly*, 27:12–22, 1956.

Lockhart, A. "The Value of the Motion Picture as an Instructional Device in Learning a Motor Skill." *Research Quarterly*, 15:181–187, 1944.

Nelson, D. O. "Effect of Slow-Motion Loop Films on the Learning of Golf." *Research Quarterly*, 29:37–45, 1958.

Penman, K. "Relative Effectiveness of Teaching Beginning Tumbling With and Without Instant Replay Videotape Recorder." *Perceptual and Motor Skills*, 28:45–46, 1969.

Penman, K., Bartz, D., and Davis, R. "Relative Effectiveness of an Instant Replay Videotape Recorder in Teaching Trampoline." *Research Quarterly*, 39:1060–1062, 1968.

Watkins, D. C. "Motion Pictures as an Aid in Correcting Baseball Batting Faults." *Research Quarterly*, 34:228–233, 1963.

Weyner, N., and Zeaman, D. "Team and Individual Performance on a Motor Learning Task." *Journal of General Psychology*, 55:127–142, 1956.

Research Reports

Markle, D. G. *Final Report: The Development of the Bell System First Aid and Personal Safety Course*. Palo Alto, Calif.: American Institutes for Research, 1967.

Newman, S. E., and Highland, R. W. "The Effectiveness of Four Instructional Methods at Different Stages of a Course." Research Report 56–58, Training Aids Research Laboratory, Air Force Personnel and Training Research Center, Chanute, Ill., 1956.

5

Behavioral Domains

Educational objectives have assumed many forms, and terminologies have tended to confuse teachers and students. This chapter includes suggestions for categorizing behaviors into four domains: cognitive, social, affective, and psychomotor. Each behavioral domain is discussed, and representative behaviors are offered. Taxonomies for the cognitive and affective domains are fairly elaborate and reasonably well structured. The psychomotor domain is still in the preliminary stages of development. Likewise, the social domain has not yet been completely ordered, but we believe it is worthy of consideration because of the potential effect of physical education on social behaviors.

Chapter Contents

The Significance of Behavioral Domains
The Psychomotor Domain
Self-Paced Activities
Externally Paced Activities
The Cognitive Domain
The Affective Domain
The Social Domain
Summary

Student Objectives

The student who understands the material in this chapter should, with the book closed, be able to:

1. Explain four values of the process of categorizing behaviors in a taxonomic form
2. Explain four types of behaviors that represent the psychomotor domain
3. Differentiate between externally paced and self-paced psychomotor skills
4. Explain the five types of behaviors that represent the cognitive domain

5. Explain the three types of behaviors that represent the affective domain
6. Explain the four types of behaviors that represent the social domain

The number and types of behaviors we can and do learn in a lifetime are impossible to tabulate. Some can be influenced even in a class situation. Researchers and theorists have devised various statistical and common-sense procedures for organizing, ordering or categorizing course or curriculum outcomes, aptitudes or abilities, response modes, or educational objectives.

One such procedure is behavioral taxonomy. The best-known taxonomy in education involves task analysis and was developed by Benjamin Bloom and his associates (see Appendix). They classified educational objectives in three behavioral domains: cognitive, affective, and psychomotor. Furthermore, they arranged the behaviors in each domain in hierarchical fashion from the simple to the complex.

In 1956 Bloom and his coworkers published *Taxonomy of Educational Objectives, Handbook I: Cognitive Domain*. David Krathwohl and others extended these efforts in *Taxonomy of Educational Objectives, Handbook II: Affective Domain* (1964). Yet to be organized is a taxonomy for the psychomotor domain, although Elizabeth Simpson prepared a plan for one, similar to the other domains, which was published in the *Illinois Teacher of Home Economics* in 1966–67.

The Significance of Behavioral Domains

Attempts have thus been made to analyze proposed educational outcomes according to their cognitive, psychomotor, and affective components. The educational objective, in terms of the behavior required, is placed in one of the domains. The behavior in turn can be further analyzed within the classification structure and evaluated at each level. By means of this process (a) the terminology used to describe educational objectives is clarified so that everyone understands them; (b) test items, examination procedures, and evaluation instruments can be easily exchanged among teachers and researchers; and (c) learning theory can be developed more beneficially. Most important, (d) teachers and curriculum builders are aided in their attempts to plan learning experiences and evaluation measures for specific objectives.

When we try to influence a particular type of behavior, that is, to encourage students to change, we must consider the particular behavioral domain in which it belongs as well as its place in the hierarchy of behaviors. Objectives emphasizing movement and coordination of body parts—e.g., learning to serve in badminton, lunge in fencing, or shoot a jump shot in basketball—are examples of psychomotor behaviors. Objectives which deal primarily with the emotions—motivation, feelings, values, attitudes, and the like—are affective behaviors. Learning to like and wanting to participate in tennis, wanting to perform well in volleyball, and becoming interested in golf are examples. Objectives involving the intellect—remembering once-learned verbal material, knowledge, interpreting, problem solving, creativity, and the like—are cognitive behaviors, exemplified by remembering rules, understanding tactics and strategies, and distinguishing between good and poor equipment.

Although behaviors are usually placed conveniently into one of three categories (psychomotor, affective, or cognitive), it would seem that a fourth category is necessary to encompass social behaviors. Physical education is one of the few educational areas that claim to affect positively such social behaviors as conduct, interpersonal relations, personal adjustment, and emotional control and stability. Students learn how to cooperate within a group to reach goals, to compete effectively but with control, to demonstrate sportsmanship, and to develop personal qualities that will encourage favorable interpersonal relations.

The identification and categorization of behaviors serve a number of purposes. A common reference system is invaluable. An analysis of behavioral components helps teachers to write objectives, to specify instructional strategies to meet these objectives, and to effectively evaluate them. This statement will become more obvious as you read.

A warning is in order here. Behaviors are categorized so that they may be dealt with easily. However, slicing all of behavior into four distinct pieces is truly not very realistic; very few behavioral expressions are in fact pure. For any behavior the major components (subjectively evaluated) determine its classification, and the domains usually interact and overlap. Athletic performances exemplify the point very well. Softball skills are termed psychomotor, and yet effective movements go far beyond mere observable execution of acts. Performance is greatly affected by both cognitive and affective behaviors (understanding and applying rules and tactics; favorable attitudes and motivation). Social behavior on the basketball court, in the form of cooperation, may also reflect skill levels, knowledge of appropriate behaviors, and a favorable inclination toward such behaviors. Understanding the behaviors that are usually associated with a particular category will help you to formulate your own class objectives more effectively and to design more appropriate testing situations.

The Psychomotor Domain

The psychomotor domain is concerned with bodily movement, control, or both. The behaviors in this domain can be characterized by the verb *doing*. When performed in a general way, they represent a movement pattern or patterns; when highly specific and task-oriented, they indicate a skill or sequence of skills. This domain includes the following kinds of behaviors, all of which could be interrelated, and any of which could be independent:

1. Contacting, manipulating, and/or moving an object
2. Controlling the body or objects, as in balancing
3. Moving and/or controlling the body or parts of the body in space in a brief timed act or sequence under predictable and/or unpredictable conditions
4. Making controlled, appropriate sequential movements (not time restricted) in a predictable and/or unpredictable and changing situation

In developing her classification schema for psychomotor domain objectives,[1] Elizabeth Simpson analyzed behaviors in vocational education, physical education, dentistry, psychology, and educational testing. As is the case with the other behavioral taxonomies, Simpson's arrangement is hierarchical, i.e., the more basic aspects of behavior precede the highly technical aspects:

Perception: the process of becoming aware of objects, qualities or relations by way of the sense organs

Set: a preparatory adjustment or readiness for a particular kind of action or experience

Guided response: assistance provided to the learner in an overt behavioral act

Mechanism: an habitual way of responding

Complex response: resolution of uncertainity and automatic performance in difficult tasks

Adaption:[2] altering responses in new situations

Origination: creating new acts or expressions

Focus

In The Physical Educator *(1970), Marvin Clein and William J. Stone applied the Simpson Scheme to the educational objectives involved in learning how to bat.*

[1] See Appendix for detailed schema.
[2] Dr. Simpson added "adaption" and "origination" a few years after developing her original plan.

"When performed in a general way, [psychomotor behaviors] represent a movement pattern or patterns; when highly specific and task-oriented, they indicate a skill or sequence of skills."

1.0 *Perception*—This is the process of becoming aware and attaching meaning to objects, events, or situations. It is interpretation, and the first link in the chain leading to motor activity. There are three subcategories indicating different levels with respect to the perceptual process.

1.1 *Sensory stimulation*—Impingement of a stimulus upon one or more of the sensory organs. The varieties of sensory stimulation are not listed in any particular order of importance. Examples are given only for those which are commonly associated with motor activity.

1.11 *Auditory*—Stimulation occurring through the sense organs of hearing. The student listens to the instructor explain how to bat.

1.12 *Visual*—Images obtained through the eyes. The student observes a demonstration of batting either through live performance, motion picture, or still pictures. Vision is the dominant sensory apparatus for many individuals.

1.13 *Tactile*—Stimulation pertaining to the sense of touch. The student is encouraged to hold the bat and feel the size, shape and texture.

1.14 *Taste*—Stimulation of the taste buds eliciting flavor in the mouth.

1.15 *Smell*—To perceive by excitation of the olfactory nerves.

1.16 *Kinesthetic*—Sensations arising in a variety of receptors located in the joints, muscles, and tendons. The learner gets the "feel" of swinging a bat. As with the preceding stimuli, kinesthetic sensations occur throughout the learning process, and are apparent in the most complex stage, automatic performance (5.2).

1.2 *Cue selection*—The process of selecting the appropriate cues to which one must respond in order to complete the task. The batter must sort out the distracting cues issued by the pitcher and "watch the ball."

1.3 *Translation*—The process of relating perception to action both during and after performing a motor act. The batter is mentally relating his movement to results and is aided in this process by feedback, especially visual and kinesthetic feedback.

2.0 *Set*—The preparatory stage or readiness to respond. This stage is best illustrated by the batter in baseball as he gets "set" for the pitch.

2.1 *Mental set*—The mental readiness to perform the task, thus mentally concentrating on the task of batting.

2.2 *Physical set*—Physically ready to perform the task by assuming the optimum posture or body position. The batter concentrates on his stance and the position of his bat prior to the pitch.

2.3 *Emotional set*—A favorable commitment toward performance, hence the batter has a desire to bat skillfully.

3.0 *Guided response*—The initial response in the development of skill and is illustrated as the student swings the bat for the first time at a thrown ball. The student may or may not receive manual guidance from the instructor.

3.1 *Imitation*—Performing an act as a direct response to the perception of another person performing it. The young batter swings the bat as he perceives the way in which the instructor demonstrated.

3.2 *Trial and error*—The performer responds in various ways, gradually refining his performance until the appropriate response is achieved. The batter eliminates extraneous movements and rejects inadequate responses until he can meet the ball with the bat.

4.0 *Mechanism*—The learned response has become habitual and the performer can now hit the ball consistently with confidence and some degree of skill. This stage assumes more complexity than the previous level and may include batting in a game situation.

5.0 *Complex overt response*—At this level the performer can accomplish a complex motor act with a high degree of skill, i.e., smoothly and efficiently. The skilled batter in athletic competition presents the best example.

5.1 *Resolution of uncertainty*—The task can be accomplished without resorting to a prior mental picture of task sequence, hence the batter can step into the batter's box and be prepared to hit without the necessity of having a mental image of the process.

5.2 *Automatic performance*—The individual can perform a finely coordinated motor skill with a great deal of ease and muscle control. The batter can now hit the ball with ease under far more complex conditions, such as, variations in the speed, direction, height, and spin of the oncoming ball. (p. 35)

Activities in physical education can be classified according to whether they are self-paced or externally paced. The instructional procedures for each are determined by the demands placed on the student.

Self-Paced Activities

Self-paced activities are those initiated by the student; he responds at his own speed to a stable object or environment. Can you think of any sport skills that would fall into this category? Write them down.

_____

_____

If you mentioned bowling, shooting a foul shot, archery, serving in tennis, gymnastics, diving, or serving a volleyball, you were on the right track. In these activities, as well as others, the student performs when ready and at his own pace. The situational cues are predictable; they do not change from moment to moment requiring instantaneous, adjusted responses. The student can preview the circumstance and respond accordingly.

When you teach these kinds of skills, you should emphasize consistency of response; quick perception of and reaction to situations are of minor importance. Self-paced skills place less of a burden on the learner's information-processing abilities, certainly a major consideration with young or very old participants. The more complex and refined self-paced skills are a great challenge to the student. They are best acquired under repetitive practice conditions.

Externally Paced Activities

In externally paced activities the situation paces the student; he must respond appropriately to unpredictable cues. It is one thing to perform well under controlled and set conditions, another to react instantaneously as the situation changes. Hence flexibility and adaptability of response are important.

Sport situations are filled with externally paced activities. Try to describe some.

_____

_____

_____

Playing a point in tennis or badminton, fencing, wrestling, playing basketball, hitting a baseball, or throwing a football are externally paced activities.

In externally paced activities, the performer must respond to the actions of his opponent. The best athletes seem to possess a vast repertoire of responses, emitted as the occasion warrants. A student learning to play basketball can often develop fine shots in practice, when he has time to shoot. But in a game he has to shoot when an opening occurs, with the opponents nearby.

Instruction in externally paced tasks should proceed from self-paced practice to the actual externally paced situation. Basics must be mastered before responses become more flexible. Heavy emphasis is placed on cue recognition and perception, for the response is only effective if properly timed both spatially and temporally.

The Cognitive Domain

The cognitive domain encompasses the student's intellectual skills and abilities as well as his knowledge and his ability to demonstrate this knowledge depending on the particular instructional objective. The most descriptive verb associated with cognition is *knowing*. Since research has indicated that a low ratio exists between what students in physical education classes know about a particular activity, as expressed on a written test, and their actual performance in it, special attention should be paid to cognitive behaviors associated with and deemed important in an activity.

Knowledges of various kinds represent cognitive behaviors, as do skills requiring comprehension, application, evaluation, and the like. The domain includes:

1. Recalling (remembering facts, ideas, or procedures)
2. Comprehending (interpreting, translating, extrapolating)
3. Analyzing (organizing patterns, relationships)
4. Solving (applying ideas, evaluating)
5. Decision making (selecting, classifying)

Benjamin Bloom and his associates were quite successful in generating ideas, research, and methods for dealing with cognitive educational behaviors. Although they were concerned with classroom behaviors rather than with physical education, certainly a number of objectives in physical education could be handled by the proposed schema. The major categories[3] of the classification are presented in hierarchical fashion:

1. Knowledge: specifics, principles, theories, methods, etc.
2. Comprehension: translation, interpretation, and extrapolation
3. Application: various circumstances
4. Analysis: elements, relationships, principles

[3] See Appendix for detailed schema.

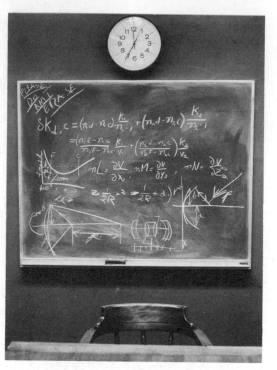

"The most descriptive verb associated with cognition is knowing."

5. Synthesis: arranging an entire structure
6. Evaluation: judgments

Cognitive domain behaviors can be part of skilled motor performance or evaluated apart from it. Most athletic endeavors require a degree of cognition in the form of perceiving relationships, analyzing and solving problems, or deciding on appropriate responses. Although the actual psychomotor act is measured in terms of response effectiveness according to predefined rules and techniques of execution, its cognitive aspects should not be taken for granted.

When measuring cognitive understandings and skills, physical educators typically give written tests, in which the recall process is most prevalent. Rules, terms, information about equipment and facilities, history, and tactics in given situations are the aspects of an activity usually tested. Sometimes written tests attempt to solicit the student's comprehension and decision-making abilities. On rare occasions analysis and problem solving are tapped. Written tests need not be the only technique for appraising cognitive behaviors, however; observation in actual or carefully simulated activities can also be effective.

Assuming the legitimacy of the hierarchy of cognitive behaviors suggested by Bloom and his colleagues, it is of interest to note that physical educators usually try to tap the lower levels of these behaviors. Tests designed to evaluate the student's "understanding" or "intellectualization" of an activity are concerned with terminology, specific facts, and the like. According to Bloom, recall is the simplest category of cognitive behaviors. Perhaps we might design better written tests or class experiences for measuring intellectual abilities and skills, which should be challenged continuously throughout the duration of the course.

The Affective Domain

Behaviors in the affective domain originate at a lower level of consciousness or awareness than other types of behaviors. At the heart of affective behaviors is emotion, or *feeling*. Interests, attitudes and values, character development, and motivation are other "internalized" processes. Although most teachers are concerned with general affective objectives, they fail to evaluate students directly, either because of the difficulty involved or because they feel such objectives cannot be graded. Furthermore, it is usually thought that affective bahaviors are slow to change, whereas psychomotor skills and informational items can be learned in a reasonable period of time.

The affective domain includes the following behaviors:

1. Valuing (selection, commitment, acceptance, preference)
2. Appreciating (evaluating, selecting)
3. Motivating (interest, persistence at)

David Krathwohl and his coworkers, who formulated the categories for the affective domain, state that these behaviors were less concrete and far more difficult to-classify than those in the cognitive domain. They were somewhat dubious about the results. The categories are hierarchically arranged on the basis of internalization.[4]

1. Receiving: willing to receive or attend to certain cues or situations
2. Responding: motivated to actively attend; willingness
3. Valuing: determining worth in a behavior, from accepting to commitment
4. Organization: structuring values in a system
5. Characterization: stabilization, consistency in value organization

[4] See Appendix for detailed schema.

Like many cognitive behaviors, affective behaviors can be evaluated as part of a skilled performance or independently of it. Certainly, the acquisition of skill and momentary fluctuations in performance can be partly attributed to interest, motivation, and attitude. These and other affective behaviors can also be determined through observation or specially designed written instruments. Thus there is no justification for the failure of teachers to state and evaluate affective instructional objectives.

Such behaviors as appreciating an activity, being interested in it, realizing its importance, persevering in it, and the like may be among the *most relevant educational objectives for the physical education class*. Working hard in class, experiencing enjoyment, and participating in activities during free time and in later years are affective behaviors that students can develop. The implication is that teachers should move away from teaching circumstances that do not encourage the formation of favorable affective behaviors toward activity. A student will spend more time practicing when he feels it is meaningful to him. The teacher may see the value of certain experiences. But the students must be cognizant of these values and evaluate the activities as beneficial or personally rewarding to them. It can be shown that participation in various activities reflects cognitive and affective behavioral components. Perhaps teachers should spend more time explaining the purpose and value of participating in activities and less time dictating them. Attempts at evaluation, as is the case with any educational objective, will reveal the teacher's success in guiding the student in the desired directions.

Krathwohl, Bloom, and Masia (1964, p. 44) have stated that "internalization is seen as related to socialization but is not a synonym for it." We feel that to do justice to the usual goals of physical education, it is more convenient to emphasize certain behaviors in the affective domain while formulating a social domain. The major categories of affective behaviors of concern to the physical education teacher are (a) interest and motivation and (b) attitudes and values. Following the guidelines of David Merrill and Irwin Goodman (1971), interest or motivation is represented by attention to or persistence at a task. "Words such as like, attend to, show an interest in, be motivated to, persist in, volunteer to, be eager to, realize the importance of, etc., are all items that suggest a motivation or interest strategy. Values can be both positive and negative. That is, some attitudes express things to avoid, while some express things to seek. Words like appreciate, see the value of, have compassion for, support, etc., often refer to attitudes or values and suggest the attitude or value strategy (p. 9)."

". . . momentary fluctuations in performance can be partly attributed to interest, motivation, and attitude."

The Social Domain

The social domain is associated with personal and social adjustment; both are related to the socialization process. Physical education is one of the few educational experiences in which students usually engage actively in situations requiring social interactions. Social interactions are effective if activities progress favorably and individual students realize personal satisfaction, development, and socially acceptable patterns of behavior. Like affective behaviors, social behaviors probably evolve slowly. Yet physical education appears to be an excellent potential medium for influencing them in the desired direction. The social domain is concerned with:

1. Conduct (sportsmanship, honesty, respect for authority)
2. Emotional stability (control, maturity)
3. Interpersonal relations (cooperation, competition)
4. Self-fulfillment (confidence, self-actualization, self-image)

When an individual becomes better adjusted personally and socially, it may be due to a variety of circumstances. But the potential of the physical education class for helping students to achieve social objectives can be somewhat realized if situations are devised

"Social interactions are effective if . . . individual students realize personal satisfaction, development, and socially acceptable patterns of behavior.

and learning experiences constructed for that purpose. Desirable social behaviors, like most learned behaviors, do not occur in a haphazard fashion but must be developed through meaningful educational experiences. Although a number of typical situations in physical education are "natural" for promoting ideal social behaviors, the teacher is responsible for using these situations to their fullest potential.

Nonacceptable cooperative behaviors can be learned as easily as acceptable ones. Nondesirable competitive actions can be learned as easily as desirable ones. The same is true for sportsmanship, emotional control and stability, self-fulfillment, and the like. Behaviors related to personal and social adjustment are important for successful sports participation as well as in a variety of everyday situations. In a sense, the demonstration of high levels of skill reflects the presence of ideally developed social behaviors. Obviously, the converse is also true. As is the case, then, with cognitive and affective behaviors, social behaviors can be observed within the context of performance in skills or apart from it.

The outstanding soccer player has learned how to cooperate with his teammates and the officials, to compete aggressively but within the established rules (under control), to display emotional stability in diverse circumstances, and to perform with confidence and a sense of personal fulfillment. If the teacher deems these and other personal and social factors important as class objectives, he should consider them when he plans the lessons for each unit. When we examine the contents and course objectives of other educational experiences offered to the student, it becomes apparent that physical education is relatively alone in its potential to influence social outcomes.

That physical education is a unique educational experience becomes more apparent as we examine the student behaviors which we attempt primarily to influence. If we can, and do, encourage behaviors thought to be worthwhile and not influenced elsewhere, our programs are quite defensible. Think about the behaviors upon which we could conceivably have impact. Various types of psychomotor, social, affective, and cognitive objectives in physical education classes might very well be possible outcomes of any physical education class. The instructor should determine those that are worthwhile and attainable.

Summary

We have attempted to classify and order behaviors for the convenience of the teacher. You should be able to identify and explain the nature of the four behavioral domains. Categories of the classification scheme for each domain are developed in a hierarchial fashion, from simple to complex behaviors. The affective domain (feeling) includes five categories: receiving, responding, valuing, organization, and characterization. The cognitive domain (knowing) includes six categories: knowledge, comprehension, application, analysis, synthesis, and evaluation. The psychomotor domain (doing) encompasses seven categories: perception, set, guided response, mechanism, complex response, adaption, and origination. As yet no categories have been established for the social domain.

Resources

Books

Education

Bloom, Benjamin S., Engelhart, Max D., Furst, Edward J., Hill, Walter H., and Krathwohl, David R. *Taxonomy of Educational Objectives, Handbook I: Cognitive Domain*. New York: McKay, 1956. Taxonomic structure of the cognitive domain, with detailed explanations.
Bloom, Benjamin S., Hastings, J. Thomas, and Madaus, George F. *Handbook on Formative and Summative Evaluation of Student Learning*. New York: McGraw-Hill, 1971. A number of chapters explain the relationship of content and behavior with regard to class or curriculum objectives in various educational areas.

Gagné, Robert M. *The Conditions of Learning.* New York: Holt, Rinehart
and Winston, 1970.   Presents classes of behavior (eight types of learning from simple to complex forms) as an alternative to the taxonomic approach offered by Bloom and his coworkers.
Krathwohl, David R., Bloom, Benjamin S., and Masia, Bertram B. *Taxonomy of Educational Objectives, Handbook II: Affective Domain.* New
York: McKay, 1964.   Taxonomic structure of the affective domain,
with complete and detailed description and examples of the categories.
Merrill, M. David, and Goodman, R. Irwin. *Instructional Strategies and
Media: A Place to Begin.* Monmouth, Ore.: Teaching Research Division,
Oregon State System of Higher Education, 1971.   Simple but clear, prac-
tical explanation of the nature of psychomotor, cognitive, and affective
domains of behavior.
National Special Media Institutes. *The Psychomotor Domain.* Washington,
D.C.: Gryphon House, 1972.   Position papers explaining different as-
pects of the psychomotor domain.

Physical Education
Singer, Robert N. (ed.). *The Psychomotor Domain: Movement Behavior.*
Philadelphia: Lea & Febiger, 1972. Contributing scholars in a variety of
educational and research areas present the scope of material encompassed in the psychomotor domain.

Articles

Education
Simpson, Elizabeth Jane. "The Classification of Educational Objectives,
Psychomotor Domain." *Illinois Teacher of Home Economics*, 10 (Winter): 110–144, 1966–1967.   Taxonomy of psychomotor domain behaviors.

Physical Education
Clein, Marvin I., and Stone, William J. "Physical Education and the Classification of Educational Objectives: Psychomotor Domain." *The Physical
Educator*, 27:34–35, 1970.   An example—learning to bat—is applied to
the Simpson organization scheme for the psychomotor domain.
Jewett, Ann E., Jones, Sue, Luneke, Sheryl M., and Robinson, Sarah M.
"Educational Change Through a Taxonomy for Writing Physical Education Objectives." *Quest*, Monograph XV:32–38, 1971.   Alternate classification scheme for Simpson's proposed psychomotor domain.

PART TWO
The Systems Approach

Part One of this text should have given you the idea that the approach we are proposing is new and somewhat different from that used in the typical methods course in physical education. We have attempted to describe the field and how it has changed over the years. We have also provided what we call a "model" of the systems approach to instruction in physical education and described a research study in which the systems approach has been used in designing instructional materials. Throughout these chapters there has been an underlying emphasis on students' behavior as the ultimate criterion of a teacher's effectiveness, as opposed to the teacher's own behavior.

Applying the systems approach as presented in this text to your own instruction will demand a commitment on your part. Much forethought must go into what is commonly called the preplanning or preactive stage of instruction. The systems approach does not require you to provide sophisticated resources, which are only available at the most fortunate schools. Rather, you must begin to consider yourself a behavioral scientist, who sets out deliberately to establish conditions which will elicit from students the specific performances and attitudes which will meet the goals of your program and curriculum. You will need to observe human behavior carefully and design measurements to indicate your successes and failures. You will need to apply data to the revision of your instructional procedures so that they will become progressively more effective.

Part Two of this book provides a detailed account of how you might implement the systems approach to instruction in physical education. We will be following, step by step, the model which was described in Chapter Three. We will now be providing more detail and ultimately more opportunities for you to become directly involved in the teaching–learning process.

6

Instructional Analysis and Performance Objectives

The major purpose of this chapter is to provide you with the knowledge and skills required to design instructional objectives which clearly communicate to both you and the student what is expected of him. However, preliminary steps must be taken to insure the successful application of the systems approach model.

Chapter Contents

Identifying Educational Goals
Classes of Educational Goals
Setting Instructional Goals

Instructional Analysis
Terminal Objectives
Instructional Analysis Process
Characteristics of a Learning Hierarchy
Benefits of Instructional Analysis
Trying Your Hand

Entry Skills, Knowledge, and Characteristics
Delimiting Entry Skills
Characteristics of Entry Behaviors

Performance Objectives
Why Performance Objectives?
Where Do Performance Objectives Come From?
Characteristics of Performance Objectives
Examples of Performance Objectives

Summary

Student Objectives

The student who understands the material in this chapter should, with the book closed, be able to:

1. Describe the various levels of educational goals
2. Write at least two instructional goals for a physical education class
3. Describe the relationship between an instructional goal and a terminal objective

4. Describe the process used in instructional analysis
5. Describe the theoretical relationship between subordinate skills and terminal objectives
6. Develop an instructional analysis for a physical education instructional goal
7. Define entry behaviors
8. Describe the relationship between entry behaviors and subordinate skills
9. List two student attributes which might be related to the achievement of an instructional goal
10. List the three major characteristics of an instructional objective
11. Indicate the derivation of performance objectives (where do they come from?)
12. Identify faulty objectives and correct them
13. Write a physical education instructional objective in the cognitive, affective, and psychomotor domains

Identifying Educational Goals

Classes of Educational Goals

In Figure 6-1, we have begun to reconstruct the systems approach model shown in Figure 3-1. At this point we are at the first step, namely, the identification of instructional goals. As we proceed, we will repeatedly show the model to remind you of the locus of the activities being described and their relationship to other components.

FIG. 6-1 *Systems approach model*

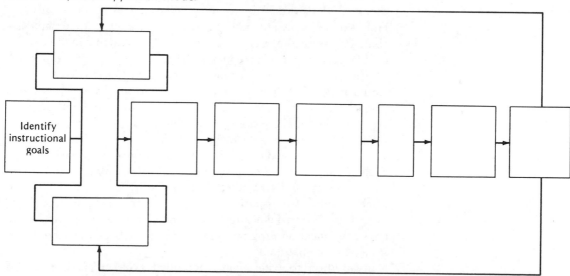

In a philosophy of education course you have probably studied the general goals of education, which have shaped and reshaped our educational system from its very beginning. In the United States the school system is basically governed at the local level. But the federal government and many state governments are tending to assume ever-expanding roles in education, so that goals may be established at any one of a number of levels. In evaluating an instructional goal, therefore, you should consider its origin and its consistency with other goals in the school system.

From time to time efforts have been made to establish educational goals at the national level. Of necessity these goals must be extremely general because they must apply to the entire United States. Such a national goal might be: "Every child shall have the opportunity to reach his peak of physical fitness." In this example there is no definition of "physical fitness" or the criteria that might be used to determine it.

Similarly, the state board of education or similar governing agency might reaffirm a national goal of physical fitness and require that each school district implement a program to achieve it. The responsibility for the most explicit goal setting often lies with the district school board. The board might indicate that 90 percent of the students in the district should meet the criteria for physical fitness as established by the President's Commission on Physical Fitness. Notice that the educational goal is much more specific than the statement from the commission.

The district curriculum supervisor for physical education or the physical education teacher could look to the commission's standards to determine the types of activities on which a student would be assessed in order to decide whether he were physically fit. From this set of criteria an integrative set of instructional activities could be designed in order to insure that the students in the district could meet this particular educational goal. As a physical education instructor it would be your task to set up such a curriculum. You would probably begin by dividing the various criteria for a student's success into such categories as running, throwing, jumping, etc., and then make more specific plans.

Thus the more remote the goal is from the teacher, the more general it usually is. For that reason he must develop very specific goals if the instructional program is to have a focus and if there is to be some basis for assessing its merit. Likewise, when the teacher looks at the specifics of a particular educational goal, he should consider whether it is consistent with other goals in the physical education program and the total goals of the school, the state, and the nation.

In addition to what might be called governmental levels of

educational goals there is a time dimension. For example, the physical fitness goals described above seem to imply that perhaps at the end of each academic year all the students in a school district would be administered a battery of tests and the results would be analyzed to determine whether, in fact, 90 percent of them did meet the commission's criteria. The goals say nothing about the pupils' fitness five or twenty five years after graduation; the emphasis is on the immediate outcome.

However, you are aware of many other educational goals which, in fact, might be considered long-range or might bring about long-lasting changes in people's behavior which could only be measured over a lifetime. For example, if we continue to use the physical fitness goal, the school board might establish a district goal that five years after graduation 90 percent of the pupils would be able to meet the commission's criteria. Certainly, it would seem that this goal would have implications for the instructional program and most certainly for its evaluation.

It is quite clear that many educational goals are not related directly to the retention of specific knowledge and skills. Many goals pertain to the development of effective citizens who can be productive in their work and can appreciate their leisure time. The individual teacher can seldom be held responsible for the achievement of these goals. Rather, it is his responsibility to design and implement a curriculum whose goals are measurable within a finite period and whose skills, knowledge, and attitudes will, hopefully, in conjunction with all the other elements of the school curriculum, produce students who meet the more general, long-term goals of the school district.

Setting Instructional Goals

As a physical education teacher your first responsibility in the instructional process is participation in the development of instructional goals. An *instructional goal* is a general statement of what a student will know or be able to do as a result of some instructional activity. Notice that the focus of the instructional goal is on the student's knowledge and ability to perform. These are behaviors you can observe and measure.

Part of your instructional goal may be a desire to instill in students a particular attitude or belief. Certainly this type of goal is legitimate; it can be derived from our broader educational goals. However, you should be cautioned that an attitudinal goal is difficult to translate into specific instructional activities. We can only provide activities and hope that certain attitudes will emerge. A common way of evaluating attitudes is simply to ask the student how he feels and what he believes. However, a much better method is to observe how he actually demonstrates knowledge or performs

". . . the focus of the instructional goal is on the student's knowledge and ability to perform. These are behaviors you can observe and measure."

in certain situations. Thus we have returned to our original statement that an instructional goal can best be stated as a description of something the student will know or be able to do. We recognize that the teacher will be working with affective and social goals, which are often difficult to achieve, and we will recommend procedures when possible.

If you were a beginning teacher and were asked to identify one or more instructional goals, where would you turn? Obviously, one answer would be the school board. From its statements of general educational goals you could probably derive one or more instructional goals related to physical education. There is another alternative—in the standard curriculum presently being used in the school you might be able to identify instructional goals either inferred or specified by the activities in which students participate. An experienced teacher might point out gaps in the present curriculum and urge that new instructional goals and supporting instructional activities be implemented to overcome them. You, as a newly trained teacher, might suggest new instructional areas for which appropriate instructional goals might be stated. As a last resort, you might turn to a college textbook.

While we have stated that the instructional goal is a general statement of something the student will know or be able to do, it should also be relatively specific. That is, it might be something the student could achieve in no less than one hour and no more than six weeks of instruction. Or, it could be a statement of one of a number of goals which might, in combination, require the entire academic year to achieve. However, the statement should probably not be simply an entry from a table of contents in a textbook. Tables of contents indicate what's in the text but not what a student should be able to do as a result of having read that text.

One possible way of determining whether an instructional goal is either too vague or too specific is to begin working with it in the context of the systems approach model. As we proceed through the model step by step, you will realize that each succeeding step depends on the product of the preceding one. By looking at your instructional goal in the context of the model, you may find that you will have to design a whole new physical education curriculum for grades K through 12. Or you may find you are dealing with instructional trivia. If either is the case, then you need to be either more or less specific in your statement of the instructional goal.

Now let's try on instructional goals for size and say that we would like to have the student be able to participate in a softball game. First of all, let's consider the basis for such a goal. The school board would probably not issue such a specific goal statement; however, it might be part of the standard curriculum. We could probably infer that participation in such a game would help build a spirit of teamwork and cooperation which might, in turn, result in the student's becoming a better citizen. Likewise, the specific set of skills required in softball could certainly facilitate better overall psychomotor coordination, which could be considered a general educational goal.

Now let's consider whether "participate in a softball game" is too big or too small a goal. This is almost impossible to determine without additional information. First, we don't really know explicitly what is meant by participating in a softball game. The student could stand in right field for nine innings and never touch the ball or come to bat several times and never take the bat off his shoulder. Clearly, we have to become more specific in terms of what our instructional goal really means.

Likewise, the magnitude of the goal will depend a great deal upon the students being instructed. If they were in fact the high school baseball team or girls' state champion softball team, then the goal would have very little meaning. However, if the class were a mixture of third- and fourth-grade boys and girls, then our task would be much greater.

These are all factors you must consider when you begin to establish instructional goals. It is impossible to provide more specific guidelines at this point for determining the proper magnitude of an instructional goal. However, your judgment will develop as you participate in this course and later when you teach.

Now write an instructional goal for a physical education activity with which you are quite familiar.

_____

_____

_____

On the basis of your instructional goal, write an educational goal which might be espoused by the school board and from which your instructional goal might be derived.

_____

_____

_____

Instructional Analysis

Terminal Objectives

In this section we will describe the process of instructional analysis. As Figure 6-2 indicates, we have now moved to the second step in the model for the design of instruction. Notice that the two boxes are connected by a solid line. This means that the statements of instructional goals are carried forward from the first box into the process of instructional analysis. These now become what we will call terminal objectives—statements of the skills you hope a student

FIG. 6-2 *Systems approach model*

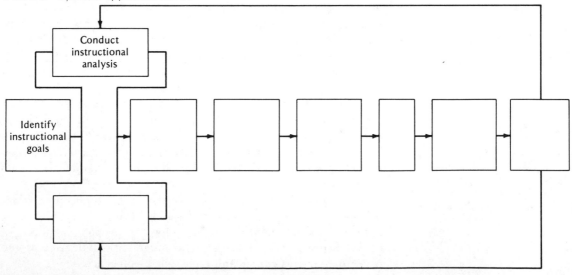

will attain as a result of your instruction. They are terminal objectives in the sense that when they are achieved the instruction will have been completed.

Instructional analysis is initiated by examining a terminal objective and determining what a student must be able to do so that, given only instructions, he can then perform that skill. The things the student must learn to perform his terminal objective are called subordinate skills. The implication of this process is that once you have identified your instructional goals, you do not necessarily consult a textbook to learn the best way to achieve that goal. In all likelihood you will not even find a textbook on the subject. When you are doing an instructional analysis, you should firmly reject the idea of simply citing the categories from a table of contents as the things that need to be taught in order to achieve that objective.

Now return to Chapter Three and restudy our discussion of the instructional analysis process. Be sure to notice particularly Figure 3-2, in which instructional analysis has been applied to the terminal objective related to baseball.

Instructional Analysis Process

Instructional analysis is like a number of sports activities: it appears to be very easy when you observe someone else doing it but is extremely difficult when you try it yourself. However, the rationale in the application of instructional analysis is based on a very straightforward psychological concept called transfer of training. When you ask, "What must a student be able to do . . .?" you are really asking what skill he should possess in order to learn another. Conversely, without that skill it would be almost impossible for him to learn the new one. Subordinate skills are almost always less complex than the terminal objective and are therefore generally easier for the student to learn. Once he has done so, there is a positive transfer to a higher, more difficult skill. When the process is repeated over and over again, the result is a hierarchy or map of subordinate skills related in such a way that if the student were to learn them in sequence, he would very probably be able to achieve the final, terminal objective; or, to state it another way, he would be maximally likely to achieve your instructional goal.

Focus

In the following passage from The Conditions of Learning *(1970),*
Robert Gagné discusses the relationship which should exist between
subordinate and superordinate skills in a learning hierarchy.

"When you ask, 'What must a student be able to . . .?' you are really asking what skill he should possess in order to learn another."

The basic functional unit of a learning hierarchy consists of a pair of intellectual skills, one subordinate to the other. As shown in [Figure 6-3], the theoretically predicted consequence of a subordinate skill that has been previously mastered is that it will facilitate the learning of the higher-level skill to which it is related. In contrast, if the subordinate skill has not been previously mastered, there will be no facilitation of the learning of the higher-level skill. This latter condition does not mean that the higher-level skill cannot be learned—only that, on the average, in the group of students for whom a topic sequence has been designed, learning will not be accomplished readily. Of course, as

FIG. 6-3 *Representation of theory regarding the functional unit of a learning hierarchy, relating the learning of an intellectual skill to the prior mastery of a subordinate skill (Gagné, 1970, p. 240)*

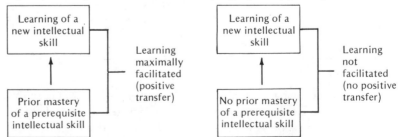

From *The Conditions of Learning*, second edition, by Robert M. Gagné. Copyright © 1965, 1970 by Holt, Rinehart and Winston, Inc. Reprinted by permission of Holt, Rinehart and Winston, Inc.

previous examples have already demonstrated, the functional unit depicted here may occur in variant forms; there may be two or three subordinate skills that contribute positive transfer to a higher-level skill. Or a single subordinate skill may facilitate the learning of more than one higher-level skill. For purposes of simplifying the discussion, however, it is convenient to consider the operation of these intellectual skills in pairs. (p. 239)

It should be pointed out that much of the developmental work with instructional analysis has been with intellectual or cognitive skills in content areas such as mathematics, science, and foreign languages. However, evidence suggests that it can be equally applied to learning tasks in the psychomotor domain.

Characteristics of a Learning Hierarchy

You may ask, "What are the essential characteristics of a learning hierarchy?" or "How would I know one if I saw one?" The first important characteristic is that each subordinate skill identified by instructional analysis logically facilitates the learning of the skill above it. If this relationship does not exist, then there is little reason to believe that learning the lower-order skill will enable the student to perform the higher-level one. For example, if we used two-skill analysis in which the superordinate skill was "The student will be able to demonstrate his ability to pitch a softball a distance of 44 feet" and the subordinate skill was "The student will be able to demonstrate how to hold a bat properly," we would have little reason to believe in the possibility of a relationship or transfer between these two skills.

At this point you may think it is difficult to tell whether the instructional analysis procedures have been applied effectively. The statements in the boxes may appear to be quite logical, but are they right? This is a good question, because there is only one way to validate a learning hierarchy, and that is to try it out empirically.

We should also note that precisely adhering to the learning hierarchy is not the only way to reach the terminal objective. It is simply the most efficient route for an entire group of students, though it may not be the optimum route for any one student.

The final characteristic of a valid learning hierarchy is that the statements in the boxes do not concern content, but rather specific skills the learner must be able to perform. Going back again to our baseball example, the verbs in the statements indicate that the student will be throwing, running, batting, catching, standing, holding, swinging, manipulating, etc.—behaviors you can observe and assess.

*Benefits of
Instructional Analysis*

A number of extremely significant outcomes are possible if you will apply the instructional analysis process to each instructional goal. First, you must think about instruction in terms of what the student will be able to do, not what you will be doing. It is your responsibility to provide the student with activities which will enable him to perform the necessary skills. In addition, you will be able to identify what might be considered the gross sequence of instruction. Note again that the instructional goal (i.e., the terminal objective) is usually a relatively complex skill which you wish the student to perform. As you move down through the learning hierarchy, you will find that the skills become less and less complex until you reach a point of very simple cognitive or psychomotor activity. Since the whole process is based on the theory of positive transfer, you should begin the instructional process at the lower levels and move successively up through higher levels until you reach the terminal objective. Here again is an avoidance of the table-of-contents approach to instruction. The hierarchy is based primarily on psychological analysis of the learning process rather than a simple content analysis.

Instructional analysis also helps to free you from using a narrowly prescribed set of contents. You may be able to identify entirely new areas of needed instruction or content topics or skills which have no valid relationship to the terminal objective. In essence, the process helps you to focus continually on the instructional goal. Finally, instructional analysis will be used in developing specific instructional objectives, designing the evaluation, and throughout the entire instructional process. In essence, it is the critical link which has been missing from many other approaches to the use of performance or behavioral objectives in the instructional process.

Trying Your Hand

As you reexamined the description of instructional analysis in Chapter Three, you noted that we began with the psychomotor abilities required in baseball. Now refer to Figure 6-4, which shows the extended application of instructional analysis to the skill "be able to bat effectively." Study the figure carefully and see if you agree with the statements of subordinate skills. Also ask yourself the question "Are the subordinate skills stated in terms of something that a student will be able to do?" As indicated above, your understanding of instructional analysis is critical because much of the remainder of the model we are describing depends on the results of such an analysis.

FIG. 6-4 *A partial learning hierarchy*

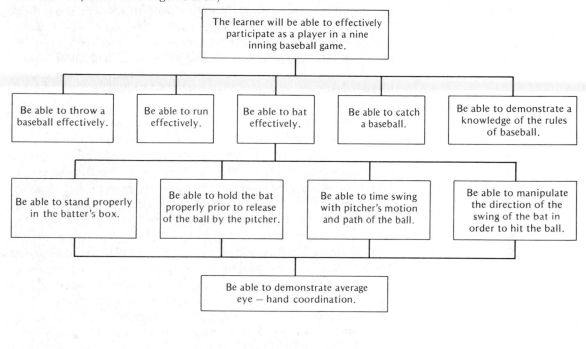

Focus

The passage below is from Robert Gagné's The Conditions of Learning *(1970). Read it carefully, substituting terminology from the field of physical education when Gagné refers to cognitive skills. See if you are as convinced as he is of the importance of instructional analysis.*

The implications for the design of instruction are clear. If learning at any level is to occur with greatest facility, careful attention must be paid to its prerequisites. It will be difficult for the child to learn the definitions (defined concepts) of geometry unless he has previously acquired the concepts of line, angle, triangle, intersection, and so on. It will be difficult for a learner to acquire the rules of any specific science unless he already knows some more basic principles of classifying, measuring, and inferring. It is demonstrably difficult for a learner to construct meaningful utterances in a foreign language unless he has learned the concept words that compose such communications; and it is difficult for him to learn these words unless he has previously learned to say the sounds of the language. Learning to read English comes hard to those who have not first learned to speak many English words. It cannot be said that any of these "shortcut" kinds of learning are impossible, because learning is a marvelous capacity of living organisms that will occur whether one wants it to or not. Nevertheless, shortcuts carry their own handicaps and typically result in deficiencies that show themselves as limitations in ability to generalize the capabilities acquired.

The fundamentals of instruction are not clearly conveyed by such expressions as "reading, writing, and arithmetic." The most general capabilities are much more fundamental than these. In subjects like mathematics and science, the most basic capabilities are to be found in the discriminations, chains, and concepts that make up the activities of observing, counting, drawing, and classifying. In languages, native and foreign, they include the basic forms of learning that characterize the acquisition of word sounds, the discrimination of these sounds, and the mastery of verbal concepts. The systematic planning of instruction in the elementary school grades in terms of such capabilities would probably have a marked positive effect on facility of learning the more advanced principles of all school subjects. (p. 274)

We hope we have convinced you of the importance of applying instructional analysis procedures to your instructional goals in order to develop a learning hierarchy. These procedures will be quite helpful to you when you attempt to identify critical subordinate skills which might be included in a learning hierarchy. Try this example, illustrated in Figure 6-5: The terminal objective in a small hierarchy is "The student will demonstrate the ability to shoot a basketball." Beneath the terminal objective are empty boxes indicating the need for at least three subordinate skills. Under one of these are two more boxes for subordinate skills. Attempt to identify skills that might be listed in the learning hierarchy by saying to yourself, "What would a student have to be able to do in order to demonstrate his ability to shoot a basketball if I simply

FIG. 6-5 *Determining subordinate basketball skills*

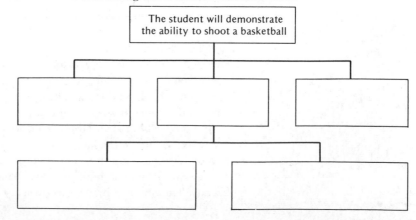

give him instructions? Would he have to be able to jump? Would he have to be able to hold the ball in his fingertips? Would he have to be able to tap the ball toward the basket? Would he have to be able to shoot with one hand? with two hands?" Enter your own statements of the subordinate skills in the empty boxes in Figure 6-4, then compare them with those of other students.

Entry Skills, Knowledge, and Characteristics

You are now ready to move to the third box in Figure 6-6 as you identify the skills the student must possess when he begins a learning activity. Notice again the focus is on the learner and his aptitude and readiness for learning. Too often we have talked about particular learning activities as being appropriate for third-grade boys or junior high students. However, these descriptors are meaningless in that they do not describe what the learners can do when they begin the instruction.

Through a number of examples we have shown that if you continue to ask the question "What must the student be able to do?" you will eventually arrive at a set of very basic and simple skills. This is true whether you are operating in the cognitive or the psychomotor domain. If the results of the instructional analysis of a complex skill were shown to a kindergarten teacher, she would probably indicate that her students would need instruction on all the skills identified in the analysis if they were to reach the terminal objective. However, if the same analysis were shown to a senior

FIG. 6-6 *Systems approach model*

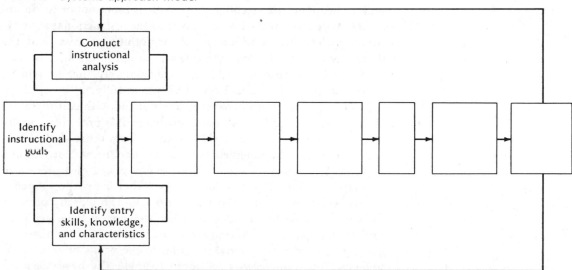

high physical education instructor, he might indicate his students already possess a number of the skills in the lower portion of the analysis. The implication, therefore, is that in these two different instructional settings with different types of students, the instructional process would begin at different points in the analysis.

As was indicated in Chapter Three, teachers have always made certain assumptions about their students' skills, knowledge, and characteristics. Rarely have they stated these assumptions specifically and tested them for accuracy.

Delimiting Entry Skills

On your instructional analysis draw a line dividing the skills you will teach from those you assume entering students will possess. All those skills above the line must be taught to the students. The question now becomes "What should be done with those skills which fall below the line?" Clearly, for those skills which fall directly below the line you could develop an evaluation instrument which would be given to all entering students. Those who demonstrated their ability to perform on this test would begin the instruction in the activity. However, those who failed to demonstrate the required entry behavior would either have to be provided with remedial activities or assigned to some other learning activity.

This consideration of prerequisite entry behaviors is extremely important if you are concerned with the success of each student in the learning process. If we assume that the theoretical underpinnings of the instructional analysis are correct—that there is a positive transfer effect from subordinate to superordinate skills, and that if the student does not possess the subordinate skills he will find it difficult, if not impossible, to develop the superordinate skills—then ignoring essential entry skills is tantamount to dooming the student to failure before he begins the instructional activity. Although learning is seldom an all-or-nothing process, and the student who lacks the requisite entry behaviors may still, in fact, benefit from the instruction, it is highly unlikely that he will be able to achieve the terminal objective successfully.

Our concern for entry behaviors does not end with consideration of the mandatory entry skills and knowledge. It is probably equally important to assess students' abilities to perform those skills which appear above the assumed entry behavior line in the instructional analysis. This assessment is, in fact, the equivalent of a pretest on the activities you will teach during a period of instruction. Such a pretest identifies not only those students who can already perform the terminal objective but many others who may have achieved some but not all of the subordinate learning skills. You can then begin to tailor the instructional program to the needs of individual students. For example, you will need to make the basic decision about using homogeneous or heterogeneous groups for instruction. If the classroom strategy is to use homogeneous groups, you will have sufficient data on the relevant skills to establish such groups. If the instructional activities seem to call for heterogeneous groups, the students can be clustered on the basis of known patterns of differences. Thus the grouping would be on the basis of specific student performance rather than on more obvious criteria such as age, sex, or body build.

A distinction should be made here between performing required entry behaviors successfully and possessing the characteristics of the target population for whom the instruction is designed. It is highly likely that we will continue to talk about instructional activities which are appropriate for college underclassmen or sophomore girls. We might even talk about explicit age ranges, such as 13 to 15, or explicit body weight, such as 150 to 180 pounds, etc. However, these qualities are related only indirectly, at best, to success in the learning process. They do, in fact, describe the general target population for whom the instruction is intended. But certainly some students who fit the description will not succeed with the instruction. Therefore, it is advantageous to you to talk about the explicit entry skills required of students who might be in the target population.

Considering entry behaviors part of the instructional process does not minimize the role of students' personality, attitudes, and general physical skills, variables which must also be included. In fact, one of the frequent primary goals of the physical education class is not only to bring about demonstrable changes in students' psychomotor skills but also to change their attitudes toward those skills. The attitudes they bring to the learning process will often influence both their learning and their attitude at the end of the instruction.

Characteristics of Entry Behaviors

Since entry behaviors are derived from the instructional analysis, they have the same characteristics as subordinate learning skills. For example, it has been hypothesized that students who have the appropriate entry skills will be aided in their learning by the positive transfer from these entry skills to the subordinate learning skills of the learning task itself. Likewise, those who do not have these entry skills will probably not succeed in the learning situation itself. Entry behaviors, like learning skills, must be validated through the instructional process. In other words, if the students who do not have what you have identified as relevant entry behaviors succeed as well as those students who do, then you must question whether you have identified appropriate entry behaviors.

Statements of required entry behaviors, like statements of subordinate skills, must focus on the actual skills the student must possess. Consequently, it would be inappropriate to list "the achievement of age 13" or "the reading level of a ninth grader" as entry behaviors. You may wish to quarrel with the statement of a ninth-grade reading level. However, there is no unanimous agreement on what a ninth-grade level is in terms of reading skill.

In order to determine whether you have understood this discussion of entry behaviors, in the space below name a subordinate skill you might wish to teach students in an area in which you are interested.

_____

_____

_____

Now write a statement of a required entry behavior for a student planning to take part in your instruction.

_____

_____

_____

If you were to show the descriptions of these two skills to a fellow student, do you think he would be able to tell you which was the required entry behavior and which the subordinate skill? He would probably assume a positive transfer relationship between the two skills. If you handed the same two statements to another student and asked him to tell you whether they described entry skills or subordinate learning skills, would he be able to tell the difference? Technically he should not be able to do so since he would not have seen the learning hierarchy or the point at which you drew the line separating the two skills.

There is an important difference between a test of entry behaviors and a pretest on the instruction which is to be conducted. The test of entry behaviors assesses those skills you assume to be prerequisite to success in your instructional program, and it is derived from those statements of subordinate skills which fall below the line you drew in the instructional analysis. The pretest, on the other hand, is an assessment of the student's ability to perform skills you will teach in your instructional program, and therefore is derived from those skills located in the upper portion of your analysis.

The entry skills test can be used to determine whether a student should be admitted to the instructional activity or whether he should first be provided with remedial instruction. The results of the pretest may be used to (a) place the students individually in the instructional program, (b) exempt them completely from that program because they already possess the necessary skills, or (c) group them (either homogeneously or heterogeneously) for more effective instruction.

Performance Objectives

Why Have Performance Objectives?

If you were to investigate other systems approach models similar to ours, you would find that many of them suggest that the teacher begin by stating performance objectives for his students. Likewise, many school districts have become aware of the need for objectives and ask their teachers to write them for all their classes. In many instances teachers attend specialized inservice institutes and workshops in which they learn to develop technically sound objectives. However, the difficulty with this approach is that no one tells the teacher on what basis or what rationale he should develop those objectives. Therefore, teachers have to rely on what they are already teaching. As a result their objectives often resemble the table of contents in a book.

If you examine Figure 6-7, you will see that the development of performance objectives for students is actually the fourth step in our systems approach model. In essence, it may be argued that per-

FIG. 6-7 *Systems approach model*

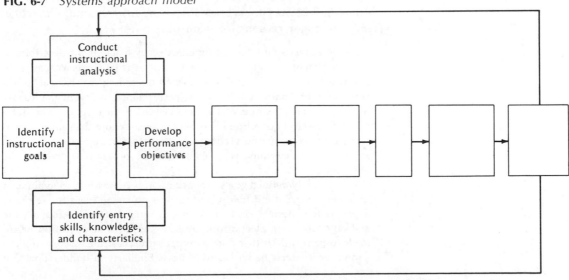

formance objectives must be developed within the context of an instructional analysis of what is to be taught, which is derived from the instructional goals established by the teacher, which in turn should be derived from the educational goals of the school district. In other words, we could argue that the teacher must be responsible for translating educational goals into instructional goals. He must analyze those instructional goals to identify subordinate skills and relevant entry behaviors before he can determine informed statements of performance outcomes for students.

The need for performance objectives became evident a number of years ago with the advent of programmed instruction. This whole movement was based on the thesis that effective learning will take place when complex behavior is broken down into its component parts and these parts are taught in small, sequential steps. As a person learns each component of the task, he should be reinforced. Through this procedure more and more complex chains of behavior can be developed until the person can perform the terminal task. This theory was incorporated into programmed instruction textbooks, in which the student reads two or three sentences, fills in a blank, moves down a page or turns it, and finds the correct answer, which should match the one he filled in. The text is designed in such a way that he is likely to have filled in the correct answer. His observation that his answer is correct is considered to be reinforcement for that behavior.

The authors of the first programmed instruction texts saw that it was critical to identify exactly what they wanted the student to be able to do when he had finished the instructional text. Robert Mager, the acknowledged originator of the concept of performance objectives, indicates in his book *Preparing Instructional Objectives* (1962) the major reasons for such objectives:

> An objective is an intent communicated by a statement describing a proposed change in a learner—a statement of what the learner is to be like when he has successfully completed a learning experience. It is a description of a pattern of behavior (performance) we want the learner to be able to demonstrate. As Dr. Paul Whitmore once put it, "The statement of objectives of a training program must denote measurable attributes observable in the graduate of the program, or otherwise it is impossible to determine whether or not the program is meeting the objectives."
>
> When clearly defined goals are lacking, it is impossible to evaluate a course or program efficiently, and there is no sound basis for selecting appropriate materials, content, or instructional methods. After all, the machinist does not select a tool until he knows what operation he intends to perform. Neither does a composer orchestrate a score until he knows what effects he wishes to achieve. Similarly, a builder does not

select his materials or specify a schedule for construction until he has his blueprints (objectives) before him. Too often, however, one hears teachers arguing the relative merits of textbooks or other aids of the classroom versus the laboratory, without ever specifying just what goal the aid or method is to assist in achieving. I cannot emphasize too strongly the point that an instructor will function in a fog of his own making until he knows just what he wants his students to be able to do at the end of instruction. (p. 3)

Experience in the use of performance objectives during the last ten years suggests that Mager was right. The teacher's "fog" will largely disappear if he bases his instructional strategies and evaluations of student performance on a clear statement of performance objectives. In addition, if the teacher is willing to share these objectives with the students, they may, for the first time, understand what they are really expected to be able to do in order to obtain a good grade and stop trying to outguess him.

Where Do Performance Objectives Come From?

In Chapter Three, as we discussed the general systems approach model, we tried to demonstrate the relationship between the statements which result from the instructional analysis process and the statements of terminal and subobjectives. This same information is included again in Figure 6-8. Look at the figure and notice that for each subordinate skill which has been identified, one or more performance objectives have been developed. For example, one of the subordinate skills in being able to participate successfully in a baseball game is "to be able to bat effectively." From this subordinate skill two subobjectives have been derived: "The learner will hit at least three of five fast balls into fair territory when the instructor serves as the pitcher" and "The learner will hit at least two of five curve balls into fair territory when the instructor serves as the pitcher."

The rationale for this direct one-to-one relationship between the performance objectives and the subordinate skills is again based on the theory that if there is positive transfer in the learning of one subskill to the next superordinate subskill, then each of these behaviors must be carefully identified and included in the instructional process. The implication is that the teacher must provide instruction for each objective and also be able to determine when the student has achieved it. Therefore, the answer to the question, "Where do performance objectives come from?" is that they come directly from the statements of the terminal objectives and subordinate skills in the instructional analysis. For each skill appearing in the instructional analysis there should be at least one performance objective which the student must achieve.

FIG. 6-8 *Relationship of instructional goal, terminal skill, subordinate skills, terminal objective, and sub-objectives*

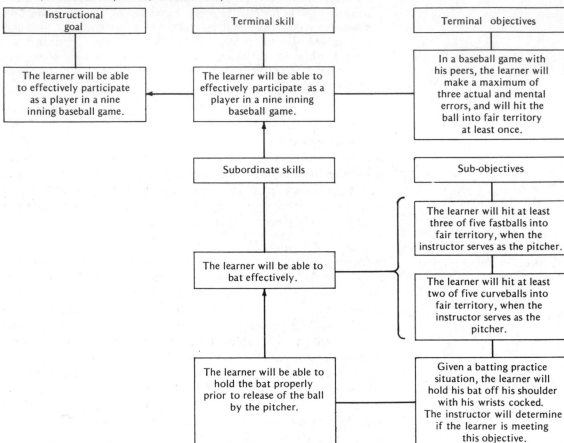

Instructional goal	Terminal skill	Terminal objectives
The learner will be able to effectively participate as a player in a nine inning baseball game.	The learner will be able to effectively participate as a player in a nine inning baseball game.	In a baseball game with his peers, the learner will make a maximum of three actual and mental errors, and will hit the ball into fair territory at least once.

Subordinate skills

Sub-objectives

The learner will be able to bat effectively.

The learner will hit at least three of five fastballs into fair territory, when the instructor serves as the pitcher.

The learner will hit at least two of five curveballs into fair territory, when the instructor serves as the pitcher.

The learner will be able to hold the bat properly prior to release of the ball by the pitcher.

Given a batting practice situation, the learner will hold his bat off his shoulder with his wrists cocked. The instructor will determine if the learner is meeting this objective.

Characteristics of Performance Objectives

You will note that we have not said much about how educational or instructional goals should be stated. We have been somewhat more specific about terminal objectives and subordinate skills, saying that they should describe behaviors the student will perform. However, performance objectives should be stated very precisely. At this point you must translate these goal and skill statements into specific descriptions of what the student must be able to do so that you, a fellow teacher, and a student would all know when you saw the behavior occurring. The purpose of this part of the chapter will be to indicate the critical characteristics of a performance objective.

Perhaps you have already been taught to write performance objectives. Let's find out how much you remember. In the space below write two objectives for an instructional area with which you are familiar.

1. _____

   _____

   _____

2. _____

   _____

   _____

Do both your objectives include specific descriptions of the behavior the student will need to demonstrate to you? Do they indicate the conditions under which he will demonstrate this behavior? Have you indicated the criterion he must meet in order to be successful? If you have included all these features, then you are probably already familiar with the development of performance objectives and may want to move ahead to the section on examples of objectives on page 133. If the objectives you developed didn't meet all these criteria, then stay with us while we discuss the characteristics of performance objectives.

First, a performance objective must clearly describe the behavior of the student when he has completed some part of the instructional process. Therefore, the subject of a performance objective statement is obviously always the student. He will be able to do something. The key word then often becomes the verb which describes what he will be doing. Educators have developed several verb lists which indicate the kinds of words which should be used in performance objectives to convey a maximum amount of meaning. Table 6-1 includes two such lists of verbs. The first includes many of the verbs associated with cognitive skills, and the second, verbs commonly associated with physical or psychomotor activity. You would obviously have little difficulty determining whether or not someone was doing these things.

However, simple verbs do not a performance objective make. It must also include a statement of the performance level expected of the student. The criterion in many performance objectives is a test score, the time required to complete a task, the distance an object must be thrown, etc. Note that all these things can be measured quite precisely. The criterion is a critical component of every performance objective because it indicates to both teacher and stu-

TABLE 6-1 *Verbs useful for making objectives more precise*

Verbal tasks	Psychomotor tasks
attend	arch
choose	bat
collect	bend
complete	carry
copy	catch
count	chase
define	climb
describe	coach
designate	coordinate
detect	critique
differentiate	float
discriminate	hop
distinguish	jump
distribute	kick
duplicate	knock
find	lift
identify	march
imitate	perform
indicate	pitch
isolate	run
label	score
list	skate
mark	ski
match	skip
name	somersault
note	stand
omit	stretch
order	strike
place	swim
point	swing
provide	throw
recall	toss
repeat	
select	
state	
tally	
tell	
underline	

Excerpted from, "Writing Higher Level Objectives: Cognitive Domain," a developmental product of the National Laboratory for Higher Education, Durham, N.C.

dent that the latter's performance is adequate and that he can move on to the next objective. If he fails to meet the criterion, then he must receive further instruction.

Though it is desirable to have objectives that can be "objectively"

"The criterion . . . indicates to both teacher and student that the latter's performance is adequate and that he can move on to the next objective."

evaluated, there are times, especially when dealing with highly skilled students, when only expert judgment can determine whether or not a student's performance is acceptable. However, as you develop objectives you should be careful not to use this criterion too often. If you do, you may well find that your objectives are ill-formed and do not really convey a great deal of information to either you or the student.

Finally, the objective must include an indication of the conditions under which the student will perform. You must indicate available resources or equipment which might facilitate his performance. For example, with a cognitive objective it is important to indicate to the student whether he may use his textbook and/or reference books in order to demonstrate a skill, or whether he must perform in a "closed-book" situation. Likewise, when you are dealing with psychomotor objectives, you should inform the student of available equipment and perhaps mention the persons he will be interacting with when he demonstrates his effectiveness. For example, in the objective shown in Figure 6-7 the student is required to demonstrate his batting efficiency by hitting balls thrown by the instructor. The objective would be quite different if the pitcher were the star performer on the college baseball team. In analyzing performance objectives, it is sometimes difficult to separate the conditions of a performance from the criteria for evaluating it. However, your primary consideration is communicating to the student precisely what you want him to be able to do.

Examples of Performance Objectives

See if you can determine whether the following performance objectives are stated properly. For example: "Standing in an open field, the student will be able to throw a baseball, on the fly, a distance of 200 feet." This would seem to be an appropriately stated objective since it describes the behavior (throwing a baseball), the conditions (standing in an open field), and the criterion (the ball must travel a distance of 200 feet).

Now, what about this objective: "The student will be able to list

90 percent of the major muscles in the body." This one isn't quite as good. The conditions under which the student will demonstrate this knowledge are not stated; you can't tell whether he must produce this list from memory or whether he can use his text as a reference.

Let's try another. "The student will be able to swim 200 yards in an olympic pool in three minutes." This one is a little more difficult to judge. It does indicate the type of behavior, namely, swimming; but it doesn't specify any particular kind of stroke. The criterion is clearly stated, namely, the student will be able to swim 200 yards in three minutes. But what about the conditions under which he will be able to do this? You know that it will be in an olympic pool, but you don't know if he will be alone in the pool or with a large number of other students attempting to perform the same objective. This could be an important distinction.

Let's try one more. "Given a standard softball field, the student will be able to hit at least three balls pitched by the instructor." The student's behavior is clearly understood (namely, hitting at least three pitches) but not the conditions under which he will perform. Will he receive three pitches or a hundred? This is a good example of an unclear distinction between the criterion and the conditions associated with an objective. You would probably consider the number of pitches the student must hit (three of ten) part of the criterion statement. In addition, this particular statement does not specify that the hits must be fair balls; according to the way it is worded, foul balls would also count.

We could go on and on with examples of this type, but you should now have an idea of the level of specificity required in a performance objective. The major point to remember is that the performance objective should be directly related to the terminal objective and the subordinate skills, which can be identified by means of instructional analysis. The performance objective should state exactly what the student will be expected to do, so that it may serve as an informal contract between you and the student. And, lastly, the statement should be sufficiently detailed that you or another teacher could determine whether the student was in fact achieving the objective.

When you completed the previous section of this chapter, you identified (page 122) five subordinate skills related to the terminal skill "The student will demonstrate his ability to shoot a basketball." Select one of those subordinate skills and in the space below write two or more performance objectives which could be derived from it.

1. _____
   _____
   _____

2. _____
   _____
   _____

Now share these performance objectives with another student to see if he can identify the behaviors, conditions, and criteria within them. Can he think of a way to evaluate the objectives?

Focus

Some educators doubt that all instructional goals can be translated into measurable statements about student performance. Others have suggested that those activities for which performance objectives can be most easily written are often the least important ones, and that the result will be an emphasis on academic trivia. The Educational Researcher, the official newsletter of the American Educational Research Association, posed these and other questions to Dr. W. James Popham, a long-time advocate of performance objectives. Participating in the discussion were Dr. C. Winston Wegner, a public school supervisor for elementary education; Mrs. Joanne Wegner, an elementary school principal, and Mrs. Tommie R. Stivers, a first-grade teacher. Part of their discussion appears here.

Dr. Popham—I think there are almost no goals, however general, which are not amenable to some form of operationalizing in order to provide the educator with better clues as to whether the goal has been achieved. No matter how exotic the subject, I think we can come close to measuring it and therefore more intelligently make subsequent decisions regarding both whether it was a worthwhile goal or good instructional sequence. I believe most objectives are susceptible to that kind of analysis by people who have the training and the time to do it. I am certainly not of the opinion that a classroom teacher can whomp up a measurable goal to handle all problems—at the end of a long day, while tripping off to the PTA meeting. That's one reason we started the Instructional Objectives Exchange. People who have the time to

Popham, W. James, *et al.* "Instructional Objectives: ER Dialogue with Researcher, Supervisor and Teacher," *Educational Researcher*, Sept. 1972, pp. 8–12. Copyright by American Educational Research Association, Washington, D.C.

work on objectives should concentrate on producing them for teachers who don't have the time.

ER—Your conviction comes through very clearly, but I have difficulty comprehending now measurable objectives can be applied, for example, to teaching an art student how to appreciate a Picasso.

Dr. Popham—One of the most difficult arenas to measure deals with internal sensations. When you talk about appreciation you identify probably the most elusive of all the things we want to develop in kids: genuine satisfaction derived from experiencing a given phenomenon. It is too easy for learners to conceal their real feelings and to display overtly deceptive information.

Still there is a way to approach this type of measurement task. Ask yourself, "How would an individual who really appreciates Picasso differ from one who really doesn't?" You can imagine a hypothetical Picasso-lover, as well as someone who doesn't like Picasso, or even art in general. Then you can imagine situations in which those two hypothetical people would *behave* differently. Such situations might relate to taking advantage of opportunities to attend Picasso exhibits, the possibility of buying a "hot" Picasso recently stolen from the National Museum of Art, or the selection of Picasso as a subject when given a chance in an English class to write an essay on any artist. There are a number of things that might distinguish between our hypothetical Picasso-lover and Picasso-hater. If you define these difference-producing situations rather clearly, you will discover some which are practical to administer. A measure of such situations, in conjunction with more prosaic self-report measuring devices, often give you a fix on whether you were promoting art appreciation.

You probably can't devise one measure which would invariably identify appreciators of Picasso, but you might find three or four measures with which to "triangulate." We spent some time a couple of summers ago generating objectives to measure kids' self-concepts and attitudes toward school. It was very hard, but it can be done. Not every measure is satisfactory but, when combined with other measures, you can make some reasonable judgments. I suppose you may discover a small proportion of nonmeasurable objectives which really couldn't be handled that way, but I doubt it.

Mrs. Stivers—Let me stay with this elusive Picasso problem a little longer. I teach first grade. I could put up a Picasso print for the children to look at, but probably the only measurable behavior that I could put down as an objective would be that they learned the names of the colors. What occurs in the affective area, if anything does, probably will happen so far distant in time that I can say little about it. You can dig up a seed right after you have planted it but you are not going to see any measurable growth.

Dr. Popham—In a way I disagree. You can't wait 20 years to decide how to teach that first grade class. Suppose you have made a grievous instructional error and are making your pupils hate art. If you wait 20 years to measure their learning, 20 classes of kids will have been damaged irreparably. The decision must be made now. I submit that there are ways of getting at some indication of the degree to which youngsters respond to art.

We have a new collection of objectives in the area of attitudes toward school and school subjects. If you are trying to promote a youngster's interest in art, you are concerned with the degree to which they are positively disposed to art. There may be no measure precisely akin to what you are looking for, but some are similar in that at the end of the program they attempt to determine whether or not the learners are willing to choose art-related activities. You might use such self-report measures. It seems to me that there are numerous ways whereby you can get a *partial* fix on whether the kids are responding positively to art during or at the end of the course.

We are going to get better at devising subtle measures in the affective domain, measures which can be administered now to allow us to make better than superficial judgments regarding the quality of an instructional program. Such measures will help us make long term predictions on the basis of short term data. A few years ago I would have made that statement with more faith than evidence, but there is some evidence now. There will soon be numerous instruments available to measure learning in the affective domain and at higher cognitive levels. Again I don't mean to imply that you will know exactly how Mary or George feels about art within five percentage points, but you will not be ignorant about how your teaching program is affecting them with respect to their reactions toward art. That kind of information, really, is very important.

Mr. Wenger—It seems to me like the immediate application of instructional objectives will be to use them with sequential skill areas, at least at the beginning. How are they most often being applied now?

Dr. Popham—I'm not necessarily concerned about the fact that many educators are currently applying measurable objectives to specific skill subjects. The extent to which educators use measurable goals in order to verify the quality of their instruction, moves them toward a consequence-orientation which is my primary advocacy. I suspect that if you get more and more educators attentive to what happens to learners as a consequence of instruction, rather than to the instructional procedures they have assembled—more attentive to outcomes than to innovations—that will be a very desirable change in the orientation of educators.

Mrs. Wenger—Let me react to that with a quotation from a paper by

Sinclair and Fischer of the University of Massachusetts for later publication. One section appears to suggest a possible danger in what you're saying now. I quote in part:

"What we identify as a key curriculum danger is analogous to Gresham's law in that an over-emphasis on behavioral objectives will drive out of our schools the various and worthwhile objectives that are not quantifiable. In outline form the argument proceeds as follows: with the continuing shortage of public funds for the schools the quest for efficiency and effectiveness, will lead to increased efforts to measure the product of the schools."

They continue to make the point that while educators may have some reservations about total reliance on measurable objectives, the public may buy this convenient new catch phrase as an educational panacea. Do you see this as a danger?

Dr. Popham—I do see very significant dangers in the fact that people tend to operationalize that which is easiest to operationalize. Many of the early examples of measurable objectives dealt with the most pedestrian kinds of consequences. One might well ask: then won't educators cleave toward the trivial and, with the emphasis on measurable objectives, won't the trivial objectives predominate? That's a possibility, but one I gladly face. I believe that most educators are sharp enough to spot trash when they see it spelled out for them. I would rather have the quality of our goals inspected in an open arena than hidden behind closed classroom doors. If the emphasis on measurable objectives yields goals which are measurable but trivial, I think educators will react quite properly by saying that this must change. The risk is there, but you have to view the risk in a trade-off context. Compare such a danger to what we currently have in education. Far too many trivial learner behaviors are being promoted behind a facade of profound goals.

Mrs. Stivers—I would like to respond to a quote from your chapter in the AERA Monograph on Instructional Objectives. In responding to a spontaneous digression from an instructional sequence, you say, "too often teachers may believe they are capitalizing on unexpected instructional opportunities in the classroom, whereas measurement of pupil growth toward any defensible criterion would demonstrate that what has happened is merely ephemeral entertainment for the pupil, temporary diversion, or some other irrelevant classroom event."

I think a lot of what we are trying to do in teaching needs time to nurture. There are things we want to foster which simply are not measurable. There are occasions when something will come up in the classrooms where the best thing to do is drop the objectives you were after and take off on that line.

Dr. Popham—I am more concerned with consequences than with prespecified intentions. Prespecified intentions help you clarify what the lesson's consequences are. Michael Scriven has an unpublished paper circulating now on *goal free evaluation* in which he says the evaluator shouldn't attend to the instructional designer's rhetoric, but rather to the consequences the instruction produces. In much the same way a teacher should be more concerned with the consequences of instruction, irrespective of whether they were planned. If you take off on some tangent and spend a week or month on it, all you have to do is verify what happened as a consequence—rather than stop in the first five minutes to conjure up an objective . . .

Mr. Wenger—One of the statements you made in the Monograph reads: "We are at the brink of a new era regarding explication of instructional goals, an era which promises to yield fantastic improvements in the quality of instruction." Do you think this prediction has been realized in terms of the amount of activity current on the subject of instructional objectives?

Dr. Popham—I would still put money on it. We will soon begin to see some of the dividends of our attention to consequence. Education in this country is such an enormously complex phenomenon that the introduction of a single factor would never be expected to have an immediate impact. The entire movement toward educational accountability seems to be related to the early furor about instructional objectives. The current concern over educational accountability would have fallen on terribly unenlightened people if educators had not earlier been attentive to USOE insistence that many federally supported programs be evaluated in terms of measurable objectives. Many educators would not have known how to go about implementing schemes for educational accountability if they did not have access to notions regarding measurable educational objectives. There are school districts right now where the attention to consequences, a focus on getting demonstrable results with kids, has yielded very significant improvements in what goes on in these schools.

Mr. Wenger—Is the trend gaining momentum?

Dr. Popham—Without suggesting that everything that's proposed under the banner of educational accountability is good, I'm very encouraged by what's happening to education in this country. There will be errors committed in the name of behavioral objectives and of accountability. Unenlightened administrators will do terrible things to their staffs by saying "you must have 25 measurable objectives by the end of 4:00 today," or something of that sort. Misuses and distortions of measurable objectives will occur in attempting to promote improvement of education. But, on balance, the overall impact will be both beneficial and, in time, dramatic.

Summary

In this chapter we have described in detail the first four components of the systems approach model. We pointed out that educational goals exist on a number of levels. The ultimate responsibility for translating these goals into more specific instructional goals lies with the teacher. Instructional goals must focus on the learner and the things which he must be able to do as a result of instruction. In order to determine those skills, we suggested that each instructional goal be translated into a statement of a terminal objective and that the required skills be identified through the use of instructional analysis. The analysis indicates the subordinate skills which, when learned in sequence from simplest to most complex, facilitate the learning of the terminal skill. We also indicated that instructional analysis is an efficient means of identifying the entry skills and knowledge students need in order to successfully participate in the instructional process. If a line is drawn through the instructional analysis at some particular point, those behaviors which fall below the line can be identified as prerequisite to entry into the learning process; those which appear above the line are skills for which instruction will be provided. Finally, we emphasized the need for performance objectives and described them. We pointed out that performance objectives become the main vehicle of communication between teacher and student as they work together to achieve the terminal objectives of the instructional program. Therefore, these statements must clearly communicate the teacher's intent and include a description of the behavior expected of the student, the criterion he must meet when performing this behavior, and the conditions under which he must perform.

We indicated at the beginning of the chapter that all these are preplanning activities, which take place well before the teacher ever encounters the student in the classroom or the playing field. While they may seem time-consuming, difficult, and sometimes frustrating, they are critical to the overall success of the teaching process, and the long-term reward in student performance is well worth the effort.

Resources

Books

Education

DeCecco, J. P. *The Psychology of Learning and Instruction*. Englewood Cliffs, N.J.: Prentice-Hall, 1968. Approximately follows the systems approach model that our book presents. Includes an excellent elaboration of entry behaviors.

Gagné, R. M. *The Conditions of Learning*. (2nd ed.) New York: Holt, Rinehart and Winston, 1970. General description and examples of the development of instructional hierarchies in a number of content areas.

Gerlach, V. S., and Ely, D. P. *Teaching and Media: A Systematic Ap-*

proach, Englewood Cliffs, N.J.: Prentice-Hall, 1971. Systems approach to teaching. Includes extensive discussion of performance objectives.

Glaser, R., "A Behavioral Science Base for Instructional Design," and Stolurow, L. M., and Davis, D., "Teaching Machines in Computer-Based Systems. In R. Glaser (ed.), *Teaching Machines and Programmed Learning, II: Data and Directions*. Washington, D.C.: National Education Association, 1965. Two of the original articles that brought the systems approach for the design of instruction to the attention of educators.

Lindvall, C. M. (ed.). *Defining Educational Objectives*. Pittsburgh: University of Pittsburgh Press, 1964. Variety of papers dealing with the definition and development of performance objectives.

Mager, R. F. *Preparing Instructional Objectives*. Palo Alto, Calif.: Fearon, 1962. Brief programmed instruction text which clearly describes the three criteria for performance objectives.

Articles

Education

Gagné, R. M., "The Acquisition of Knowledge," *Psychology Review*, 69:355–365, 1962; Gagné, R. M., and Paradise, M. E., "Abilities and Learning Sets in Knowledge Acquisition," *Psychological Monographs*, 75 (no. 518), 1961. Gagné's original articles describing the experimental research that supports the notion of instructional hierarchies.

Gagné, R. M. "Learning Hierarchies." *Educational Psychologist*, 6:1–6, 1968. Gagné's observations about learning hierarchies seven years after his original research.

"Instructional Objectives: An ER Dialogue with Researcher, Supervisor, and Teacher." *Educational Researcher*, 1 (no. 9):8–12, 1972. Complete text of interview with James Popham.

Physical Education

Davis, Robert. "Writing Behavioral Objectives." *Journal of Health, Physical Education, and Recreation*, 44:47–49, 1973. Describes the writing of objectives specifically for physical education.

Shockley, Joe M. "Needed: Behavioral Objectives in Physical Education." *Journal of Health, Physical Education, and Recreation*, 44:44–46, 1973. Explains the nature of instructional objectives and their physical education implications.

7
Evaluation

Instructional evaluation is the process of collecting and analyzing data in order to determine the degree to which prestated goals have been achieved. If the purpose of instruction is to facilitate positive changes in student behavior, then obviously the purpose of evaluation is to assess those changes, and in turn, the effectiveness of the instruction. In the process of systematically examining the product, student learning, we lay the foundation not only for grading and diagnosing the student but also for evaluating the process, the actual instruction. In other words, the effectiveness of the instruction is indicated by the degree to which students achieve their objectives.

In this chapter we will present several major concepts that are critical to the development of a systematic approach to teaching. Admittedly, the use of evaluation is limited in much the same way instruction is. That is, you may have excellent ideas about how to teach a particular activity, but if the proper equipment is not available, you must adopt an alternative procedure. Likewise, you may have to reconsider an ideal (but not idealistic) evaluation plan because of a lack of media or an abundance of students. The major point to remember is that the concept of criterion-referenced testing we are about to discuss is equally valid in both good and not so good instructional situations.

This chapter describes a rational approach to the evaluation of students and instructional activities. It includes criterion-referenced testing, the process of which is based on prestated performance objectives, and instructional evaluation, which is based on a prestated criterion of instructional effectiveness.

Chapter Contents

Criterion-Referenced Testing
Designing Criterion-Referenced Tests
Examples of Criterion-Referenced Tests
Norm- vs. Criterion-Referenced Evaluation

Instructional Evaluation
Setting an Instructional Effectiveness Criterion
Phases of Instructional Evaluation
Components of Instructional Evaluation
Summary

Student Objectives

The student who understands the material in this chapter should, with the book closed, be able to:

1. Identify and define the three major characteristics of a criterion-referenced test item
2. Construct test items, or devise assessment techniques, for objectives in each of the four domains of learning
3. Identify and define the two major criteria of a well-constructed criterion-referenced test
4. Develop a criterion-referenced test
5. List four types of criterion-referenced tests and their purposes
6. Formulate at least two different grades which could be assigned to a student based on the results of a criterion-referenced test
7. Construct and summarize an objective-achievement matrix (as the basis for instructional evaluation)
8. List questions which would be appropriate for an instrument designed to assess student attitudes and opinions concerning the instructional process

Criterion-Referenced Testing

Criterion-referenced testing is the process of developing and administering assessment instruments based directly on performance objectives in order to determine whether or not students have achieved those objectives. Results of such testing form the basis for assigning student grades and permit the identification of specific areas of difficulty for each student so that future instruction can be appropriately prescribed. Figure 7-1 illustrates the function of criterion-referenced testing within the systems model.

Designing Criterion-Referenced Tests

Designing a criterion-referenced test involves developing test items and assembling them into an assessment instrument. The nature of an item is wholly a function of the objective on which it is based. Objectives in the cognitive domain generally require paper-and-pencil assessment items or items which call for a specific product or performance. Generally, it is relatively easy to determine achievement of a cognitive objective; either the student "knows" the appropriate response or he doesn't. Assessment in the affective domain is not quite as simple. Affective objectives are generally

FIG. 7-1 *Systems approach model*

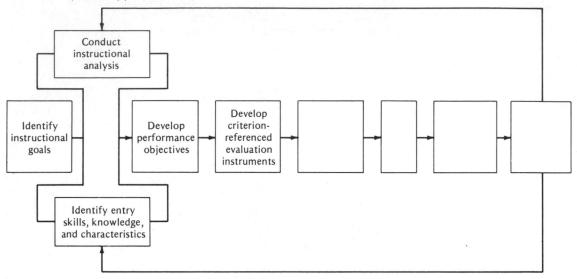

concerned with the student's attitudes or preferences. As there is no way to directly measure a student's attitude (e.g., whether he enjoys baseball), items for affective objectives generally require either that the student state his preferences or that the teacher observe the student's behavior. For example, if a student voluntarily engaged in baseball games on three different occasions, you might infer that he enjoys baseball.

Items for objectives in the psychomotor domain are most like items in the cognitive domain, which require demonstration of a specific performance. While an objective of this type in the cognitive domain might require a student to demonstrate his ability to make an oral presentation which met prespecified criteria, an objective in the psychomotor domain might require him to demonstrate his ability to properly execute a chip shot in golf, again according to prespecified criteria. Similarly, items for objectives in the social domain are most like items in the affective domain. There is no direct way to determine whether a student is a "good sport." As with affective objectives, the assessment of social objectives relies primarily on observation of student behavior and, to a lesser degree, on students' reports. For example, if a student accepts the referee's decisions without question, especially when they are to his disadvantage, you infer that he is a "good sport."

Regardless of the educational domain, however, a good criterion-referenced test item must possess several characteristics. The most important of these is congruence with the objective, i.e., the response required by the item must correspond directly to the desired student behavior as stated in the objective. If an objective stated that a student should be able to execute a tennis serve properly, then an item asking the student to name three types of serves would not be appropriate. The only acceptable "test" would be one that required the instructor to observe the student executing a serve. If indeed knowledge of types of tennis serves were an identified subordinate skill, then an additional objective would be required specifying that the student be able to name at least three of them.

Another important characteristic of an item is clarity. An item should be stated clearly enough so that the learner understands exactly what he is to do. It should also be stated in such a way that any competent evaluator could observe a student and make a definite decision as to whether he was making the appropriate response. If an item stated that the student should "write a one-page essay on bowling etiquette," it would probably not be clear to either the student or an evaluator (other than the person who wrote the item) whether the student was to discuss the importance of etiquette, its rules, or some other unidentified topic. If, however, the item stated that the student should "list at least seven rules of bowling etiquette as stated in the Bowling Handbook," the correct response would probably be obvious.

A third characteristic of a criterion-referenced item is sufficiency, in that an acceptable response to it constitutes sufficient evidence that the intended objective has been achieved. In other words, it is relatively easy to determine by observing the behavior whether it meets the criterion as stated in the objective. In order to clarify this concept, we will examine an objective and three possible items:

Objective: Given a bowling score sheet which indicates how many pins were knocked down in each of ten frames, the student will correctly compute and insert the bowler's score in each of the ten frames.

Possible Items:

1. On the score sheet that follows, compute and insert the bowler's score in each of the ten frames.

6	2	7	1	5	4	3	3	9	–	F	8	7	2	6	3	6	1	8	1

2. On the score sheet that follows, compute and insert the bowler's score in each of the ten frames.

"The response required by the item must correspond directly to the desired student behavior as stated in the objective."

8	1	6	◣	3	6	◼	9	–	7	◣	8	1	7	2	9	◣	6	3	

3. On the score sheet that follows, compute and insert the bowler's score in each of the ten frames.

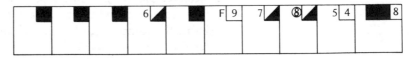

◼		◼		6	◣	◼	F	9	7	◣	⑧	◣	5	4	◼	8

Which of these items do you think best measures the objective? Why?

_____

_____

You would probably agree that even though Item 2 is better than Item 1, Item 3 is the most acceptable because it best represents the range of possible frame scores. Item 1 tells us nothing about the student's ability to score spares and strikes, and Item 2 tells us nothing about his ability to deal with complex combinations of strikes and spares. These two items would have been clearly unacceptable if the objective had been stated as follows:

Given a bowling score sheet which indicates how many pins were knocked down in each of ten frames, *and which includes examples of open frames, spare frames, strike frames, and combinations of these*, the student will correctly compute the bowler's score and insert it in each of the ten frames.

Examples

By now you should be reasonably knowledgeable about the types and characteristics of criterion-referenced test items. At this point we will look at a sample objective from each of the four learning domains and items appropriate for each objective. You will be given an opportunity to construct an item for each domain and to compare it to the appropriate sample item. It is important that you become proficient in writing items in the four domains since physical education is one of the few areas in education (possibly the only one!) in which attempts are made to foster the achievement of objectives in all of them.

The Cognitive Domain

Objective: Given a list of rule violations, the student will correctly identify from a second list the team sport associated with each.

Sample Item: For each of the terms in Column I, find the name of the associated sport in Column II and write its letter in the blank in front of the appropriate term.

Column I

_____ 1. Error
_____ 2. Holding
_____ 3. Goal tending
_____ 4. Icing

Column II

_____ A. Football
_____ B. Hockey
_____ C. Baseball
_____ D. Basketball

Objective: Given a list of final scores (the scores at the termination of the game), the student will correctly indicate whether they were in baseball, football, or basketball.

(In the space below construct a criterion-referenced test item for the preceding objective.)

Sample Item: _____

_____

_____

_____

Sample Item: Below is a list of final scores. For each score indicate with an A, B, or C which sport probably resulted in that score.

A = Baseball
B = Football
C = Basketball

_____1. Browns 120—Blacks 105
_____2. Bluebirds 2—Hummingbirds 1
_____3. Cougars 21—Panthers 17.
_____4. Angels 12—Devils 0
_____5. Giants 77—Midgets 60
_____6. Gorillas 42—Orangutans 21

The Affective Domain

Objective: Given a choice, the student will engage in a physical rather than a sedentary activity.

Sample Item: For each of the following pairs of activities indicate with an X which you would rather do.

1. Watch television _____ Take a walk _____
2. Play golf_____ Play bridge _____
3. Read a book _____ Go bowling _____
4. Play scrabble _____ Go on a bike hike _____
5. Take a nap _____ Go swimming _____

Objective: The student will demonstrate his preference for the game of football by participating passively as a spectator and actively as a player.

Sample Item: _____

_____

_____

_____

Sample Item: Please answer the following questions as accurately as you can in the space provided.

1. Did you try out for the school football team?

_____

2. How often do you watch professional football games on television?

3. Have you joined the school pep club?

4. How often do you play football with your friends?

Psychomotor Domain

Objective: The student will successfully make six of ten free throws on a standard basketball court.

Sample Item: Eight students are assigned to a basket. Three stand on either side of it on the appropriate lines; one is the shooter, and one marks the evaluation sheet. After each student has shot his ten shots, the group rotates one position.

Evaluation Sheet (Mark X if foul is made, 0 if missed.)

Name	1	2	3	4	5	6	7	8	9	10
John Doe	X	0	0	X	X	0	X	0	X	X
Jim Smith	0	0	X	X	X	X	0	X	X	X

Objective: The student will develop and perform an original floor exercise routine which includes at least three distinct, completely executed movements.

Sample Item: _____

_____

_____

_____

Sample Item: Instructor to Student: "Please perform your routine. Each movement must be completely executed to receive credit. For example, if you cannot recover from a back bend, you will not receive credit for it. Begin when you are ready."

Movement Attempted	*Completed*	
1. _____	Yes	No
2. _____	Yes	No
3. _____	Yes	No
4. _____	Yes	No

etc.

"Objective: The student will successfully make six of ten free throws on a standard basketball court."

Social Domain

Objective: The student will demonstrate good conduct in his response to the referee's calls and in his behavior toward the opposing team.

Sample Item: The student is to be observed while he is playing in at least two games and a record made of his responses in the following situations.

1. How many referee decisions were unfavorable to the student? _____

2. How many times did he question unfavorable decisions? _____

3. How many times did he swear or use abusive language when speaking to an opponent? _____

4. If the opponents won the game, did he congratulate them? _____

5. If his team won the game, did he make a positive statement to the opponents, such as, "You played a good game"? _____

Objective: In a basketball game the student will demonstrate cooperation by passing rather than shooting when a team member is in a better position to shoot.

Sample Item: _____

_____

_____

Sample Item: The student is to be observed during at least one entire game. A record is to be made of the number of times he has an opportunity to shoot, how many times he actually does so, and how many times he passes the ball to a team member who is in a better position.

By now you are surely ready to develop an acceptable test item for almost any objective. You should understand, however, that the task may be more difficult for some objectives than for others. Occasionally you may have an objective for which you can devise no acceptable item. One possibility in such a situation is that the objective is too large, in which case it may need to be broken down into several more specific objectives. Another, less probable, possibility is that the objective, though worthwhile, is simply not measurable. If it is not, you should be sure the students understand that they will not be responsible for achieving it.

Can you think of at least one objective you might include in your instruction for which you can devise no acceptable test item?

_____

_____

_____

Types of Criterion-
Referenced Tests

After you have developed your test items, the task remains of assembling them into several different types of tests for use at different stages of instruction. First, you will need a test to measure the achievement of objectives assumed to have been met in previous instruction, i.e., a test of entry skills and knowledge. More accurately, the test of entry behaviors should measure those skills prerequisite to achievement of the objectives you have selected for your instruction. Second, you will need to develop a pretest to determine whether any of your students have already achieved some of those objectives. You may often find that some students have already achieved one or more of them. In most cases you will probably want to exempt these students from that portion of the instruction that is concerned with those objectives. The test of entry behaviors and the pretest are both diagnostic in that future activities may be prescribed on the basis of students' performance on these tests. This information is also of great importance in the revision of the instructional activity.

Third, you will need a posttest to assess the achievement of objectives following instruction. While the posttest can also be considered a diagnostic instrument, its primary purpose in relation to student evaluation is to form the basis for the assignment of grades. As will be discussed later in the chapter, analysis of posttest scores also forms the foundation for instructional evaluation. Lastly, a retention test may be constructed to measure the achievement of objectives after an interval of time has elapsed since instruction. The permanence of a mastered skill is important in that the achievement of later objectives may depend on it. Analyzing the pattern of objective retention provides further input to the instructional evaluation process.

As you may have already realized, all, or a portion, of one type of test may be identical to all, or a portion, of another. The degree of overlap is primarily a function of the level of the objective. An objective at the knowledge level, such as ". . . should be able to list four rules of baseball," can probably be measured by the same item on the pretest, posttest, and retention test. An objective at a higher level, however, such as ". . . should be able to correctly compute the bowler's score," should probably be assessed with different, but equivalent, items on the three tests, primarily to insure that achievement is based on ability to perform the task and not on recall of a previous one. Therefore, the nature of the objective will determine the number of items which must be constructed to measure it.

As in the construction of items, a number of factors should be considered in the construction of a test, regardless of its purpose. Two primary requirements are validity and reliability. In order to be valid, a criterion-referenced test must measure the objectives it purports to measure. If all the items have the characteristics previously discussed, namely, congruence with the objective, clarity, and sufficiency, they will be valid, and hence the resulting test will be valid. Test reliability is generally indicated by the consistency or stability of test scores. Success or failure on one item or set of items should indicate success or failure on another item or set of items within the same test; achievement on a test should be repeated if the same test is readministered. Conversely, performance on an unreliable test gives no indication as to how the student would perform should the test be given again. The best way to insure reliability, however, is to develop a valid test; a test that measures what it is intended to measure will tend to do so consistently. There are a number of statistical procedures for computing reliability. However, they are based on the assumption that scores will vary, which makes them generally inappropriate in a system whose goal is the achievement of a given level of performance by all students.

Several other factors should be considered in developing a paper-and-pencil test. First, group together items dealing with related objectives. Also group together, with a common set of directions, items in the same format (true-false, multiple choice, completion, etc.). State the directions very clearly; the student should know exactly what he is to do and by what criteria his performance will be judged. Lastly, to avoid confusing students, do not divide an item between two pages. Though there are a number of guidelines specific to particular item formats (e.g., completion items), discussion of these is beyond the scope of this book.

You should now be able to construct criterion-referenced test items and assemble them into an appropriate test. Briefly, a criterion-referenced test item should be objective and communicable to students and evaluators. It should also be a sufficient indicator of objective achievement. Criterion-referenced test items can be assembled into tests of *entry skills and knowledge*, which measure assumed competencies; *pretests*, used to determine which instructional objectives have been achieved prior to instruction; *posttests*, designed to assess achievement of objectives following instruction; and *retention tests*, which yield an index of the permanence of learning. A well-constructed test is valid (i.e., it measures what it claims to measure) and reliable (i.e., it measures consistently).

Norm vs. Criterion-Referenced Evaluation

Evaluation, or grading, means representing a measure of students' achievement by means of a symbol. Grading based on criterion-referenced test results is known as criterion-referenced grading. Traditional grading, however, indicates how well a student has performed *in relation to others in the class* and is known as norm-referenced grading.

In a norm-referenced system, an attempt is made to devise tests which will result in a "spread" or wide range of scores. Grading is typically based on a normal curve approach; that is, the majority of students receive C's, those with the very highest scores receive A's, those with the very lowest scores receive F's, and so forth. A grade in such a system tells us little about what a student actually knows or can do. An A tells us only that a student did better than most of the others in his group. A norm-referenced system is generally group-oriented; all students progress from topic to topic

"A norm-referenced system is generally group-oriented; all students progress from topic to topic at the same rate."

at the same rate. To a great extent a student's grade reflects his ability to keep up with the class. As a consequence, many students move on to succeeding units without having mastered the important concepts or skills in the previous unit. As the earlier concepts may be prerequisite to the new ones, these students are at a distinct disadvantage before they even begin. The result is a "snowball" effect, with some students falling progressively further behind as time passes.

A criterion-referenced system is in many ways a mirror image of a norm-referenced system. In the former a wide range of student scores on a posttest indicates that the instructional system has failed to achieve its goal, namely, that most students achieve most, or all, of the instructional objectives. Criterion-referenced grading is based on an absolute rather than a relative standard. Usually, if a student achieves the required number of objectives he receives a "pass," if not, a "fail." The pass–fail decision is made individually for each student and is based on his performance, without any reference to the performances of others in the class. A grade of pass in a criterion-referenced system, in contrast to an A in a norm-referenced system, can be translated into specific information, i.e., what the student knows and what he can do.

Another major difference between the two grading systems is that a criterion-referenced system is generally individual-oriented; students do not necessarily progress from topic to topic at the same rate. A student's grade reflects his ability to exhibit the required behaviors, rather than his learning rate. Rate of progress is an important concept in that some students require more time to achieve a given set of objectives than others. In a pass–fail system the proportion of students who receive a fail is generally very small. Each student is allowed to progress at his own rate and to take a test when he is "ready." Criterion-referenced testing helps to reduce failure even further. A teacher can quickly identify areas of weakness and prescribe appropriate remedial instruction before the student progresses to a new unit. Thus all students who begin a new unit will possess the same knowledges and skills required for that unit.

The weaknesses of the norm-referenced grading system and the strengths of the criterion-referenced system suggest that the ideal system would contain no grades at all; a record would simply be maintained indicating objectives achieved to date by each student. Such a system would eliminate the concept of failure from grading, and reporting pupil progress would be greatly facilitated. A report card would consist not of letters but of a list of achieved objectives. Similarly, if a student's progress were compared with that of others in his group (if for no other reason than because parents generally desire this type of information), such a comparison could be made in terms of objectives achieved. Such an ideal system would have as one primary goal the achievement by all students of those objectives considered to be minimum essentials. Promotion from one grade level to the next (if such existed in our ideal system!) would indicate demonstration of at least a specified set of minimum skills and knowledges.

Unfortunately, the policies and procedures of most educational systems do not permit the implementation of our ideal system at this time. Teachers must manage as best they can within the existing structure. One common problem is that often a teacher who is operating a criterion-referenced pass–fail system must at some point produce a letter grade for each student. This situation can be handled in several ways. One is to assign letter grades on the basis of the number of objectives achieved within a given time interval. As you can see, this approach greatly resembles norm-referenced grading. Another, more satisfactory approach involves the use of a performance contract, which is an agreement between student and teacher concerning the objectives to be achieved in a specified period of time. The teacher's role is to guide the student in formulating an agreement which is appropriate to his unique set of aptitudes and abilities. Using performance contracting the teacher can assign a letter grade based on actual, rather than expected, student performance. Obviously, student motivation is a major determinant of such a grade. The important point is that the student is graded on the basis of his own achievement, not in relation to others in the class.

There is a definite trend in education toward adopting a criterion-referenced grading system. Increasingly, instruction is being based on prespecified objectives, and the evaluation of students is dependent on the achievement of those objectives. New grading systems generally result in pass–fail or credit–no-credit assessment. Some systems also permit the achievement of honors–pass or honors–credit. Whether or not you will be teaching in a school which encourages or permits criterion-referenced grading, this section has described several methods by which you can assign grades (in any form) based on individual achievement of objectives.

Instructional Evaluation

Instructional evaluation is the process of assessing instructional goals. It can be used prior to, during, and following instruction. Basically, instructional evaluation involves examining instructional content and procedures in order to explain and rectify discrepancies between instructional goals and actual student performance. The purpose of this section is to identify the type of data needed in order to revise instruction. Chapter Eleven will indicate how such data are actually utilized.

Setting an Instructional Effectiveness Criterion

An instructional effectiveness criterion is the overall level of student performance desired for a given set of objectives. Such terms as 90/90 or 80/80 are commonly used. A 90/90 criterion, for example, simply means that if the instruction is effective, 90 percent or more of the students will achieve 90 percent or more of the objectives. An individual student would need to achieve 90 percent of the objectives in order to receive a pass; 90 percent of all the students, however, would have to achieve 90 percent in order for the instructional activity to be judged effective. The teacher sets the actual criterion level, based on the nature of the objectives. In some cases, if the objectives to be achieved would be required for a variety of subsequent learnings, a 95/95 criterion might be most appropriate. Usually, however, it would be unrealistic to set a 100/100 criterion.

Analysis of posttest scores reveals the degree to which the criterion has been reached. A large discrepancy between desired and actual achievement generally suggests that considerable revision of the instructional activity is needed, whereas a small discrepancy will probably require only minor revision. The purpose of instructional evaluation is to identify the weaknesses in the instructional activity which contributed to the lack of achievement. This process is feasible only in a criterion-referenced system; objectives not achieved can be quickly identified and that portion of the instruction associated with those objectives pinpointed for examination. As the above discussion suggests, instructional evaluation depends heavily on posttest scores. However, the entry behavior and pretest scores, along with student attitudes and comments and teacher observations, are also extremely valuable.

Phases of Instructional Evaluation

You will usually analyze the instructional process at three different times: prior to, during, and immediately following instruction. Analysis yields different information at each stage. Before using materials and procedures with large numbers of students, it is generally good practice to try them out with a small number of students in order to identify major problems or glaring errors. You can then revise them immediately, on the basis of student feedback (in the form of written or oral comments) rather than student achievement.

Instructional analysis in the remaining stages is based primarily on student achievement. The reasons for lack of achievement are found by analyzing instructional content and procedure, a process which relies on both achievement data and student feedback data. Analysis of within-learning data and posttest data enables you to determine the permanence of learned skills, if indeed they were learned. You might find that some objectives were never achieved by some students. If objectives are not achieved during learning, you can normally not expect performance on a posttest. If, however, some objectives are achieved during learning but are not achieved on a posttest, you will draw different conclusions concerning the adequacy of instruction.

Components of Instructional Evaluation

Regardless of the phase involved (except for the first, which is prior to instruction), instructional evaluation involves two basic components, analysis of instructional content and analysis of instructional procedures. The former is based primarily on analysis of achievement data, the latter on student feedback and teacher observation data.

Content analysis is based on the objective achievement rate, which is determined by dividing the number of students who achieve an objective by the number of students who attempt that objective. Objective achievement analysis can be facilitated by means of a matrix such as the one shown in Figure 7–2 (a 1 indicates achievement of the objective). Ideally every objective should have an achievement rate of 4/4 or 1.0. The closer the achievement rate is to 0.00 (in this case 0/4), the more suspect the related instruction becomes. Of course, the fault may lie in the wording of the objec-

FIG. 7-2 *Objective achievement matrix*

	Obj. 1	Obj. 2	Obj. 3	Obj. 4	Obj. 5	Obj. 6	Obj. 7
Student 1	1	1	1	0	1	0	1
Student 2	1	1	1	0	0	1	1
Student 3	1	1	1	0	1	0	1
Student 4	1	1	0	0	0	0	1
Error rate	1.0	1.0	.75	.00	.50	.25	1.0

tive or in the construction of the criterion-referenced test item; these possibilities should also be investigated. (This subject will be discussed in more detail in Chapter Eleven.)

Analysis of student feedback data is an important part of instructional evaluation. Such data can be collected in many ways. Can you think of at least two?

1. _____

2. _____

Probably the two major methods are the structured interview and the attitude questionnaire. If you named at least one of these, then you are beginning to think like an evaluator! A structured interview is usually conducted following the administration of a posttest. Questions are carefully planned in advance and are designed to elicit information pertinent to the effectiveness of instructional procedures.

Whether the interview is conducted orally or is in the form of a questionnaire, it will generally contain both fixed-alternative and open-ended items. Fixed-alternative items ask the student to select his response from a prespecified set of alternatives. Open-ended questions provide only a frame of reference; the form of the response is entirely up to the respondent. The example below contains both a fixed-alternative and an open-ended item.

Example: How effective do you feel the films were in enhancing your understanding of the instruction?

a. _____ Very effective
b. _____ Effective
c. _____ Depended on the objective
d. _____ Not very effective
e. _____ Very ineffective

If you selected a, b, or c, which film (or films) was most effective, in your opinion?

_____

Why? _____

_____

Can you list some other questions which might be asked in such an interview?

_____

_____

Some additional questions might be:

Did you understand exactly what each of the items on the posttest was asking you to do? Yes _____ No _____ If not, with which items did you experience difficulty?

_____   # _____

_____   # _____

_____   # _____

Did you feel that the instruction was sufficient for each of the instructional objectives? Yes _____ No _____ If not, for which objectives was the instruction insufficient?

_____

_____

If you were going to revise the instruction, what changes would you make?

_____

_____

What features of the instruction would you retain?

_____

_____

Though a structured interview will probably include attitudinal questions (e.g., did you _feel_ the instruction was adequate?), an attitude questionnaire is probably more helpful in determining how the students reacted emotionally to the instruction. Did they enjoy it? Did they find it dull? Instruction can be entirely adequate and very boring at the same time! In constructing an attitude questionnaire, you could use item formats similar to those included in the structured interview, or you might wish to include scale items. Such items ask students to place their attitude somewhere on a continuum.

Example: Please place an "X" on the line in the interval which best represents your attitude.

The written materials on the history of tennis were:

1	2	3	4	5

Very
uninteresting

Very
interesting

or

The written materials on the history of tennis were interesting.

1	2	3	4	5

Strongly
disagree

Strongly
agree

Can you think of at least two other questions which might be included in an attitude questionnaire designed to collect information on students' feelings about the instruction?

1. _____

2. _____

If you listed questions concerned with student interest, preference, or disposition, your thinking is in the right direction.

You should now be familiar with, and be able to use, several procedures and tools for collecting information on the effectiveness of instruction and for subsequently revising it: objective achievement analysis, the structured interview, and the attitude questionnaire. A criterion-referenced test provides sufficient data for student grading but tells us little about the relative effectiveness of our content and procedures as reflected by analysis of objective achievement and as perceived by students. In Chapter Eleven we will discuss in more detail the actual revision process based on these data.

Summary

Instructional evaluation involves assessing students' progress as well as instructional content. Student evaluation is based on the results of criterion-referenced tests, which in turn are based on performance objectives. The purpose of student evaluation may be to diagnose, as with the test of entry behavior and the pretest; or to assess achievement, as with the posttest and retention test. Whatever the purpose, test items must be clearly stated and directly related to the objectives. Appropriate items must then be assembled into valid, reliable test measures. The key to developing a good test is to develop good items. Grades based on test scores are usually assigned on a pass–fail basis in a criterion-referenced system, although other methods are possible. Regardless of the symbols used, grades should be based on individual achievement, not group progress.

Instructional evaluation determines the effectiveness of instruction. Within-learning and posttest data yield information concerning the effectiveness of the instructional content. Specific problem areas can be readily identified through the use of objective achievement analysis. Student feedback, collected in structured interviews and attitude questionnaires, can reveal reasons for ineffective instruction.

Continuous program revision and improvement are integral components of the systems approach to instruction. Careful identification, collection, and analysis of data are the key to effective revision.

Resources

Books

Education

Bloom, B. S., Hastings, J. T., and Madaus, G. F. *Handbook on Formative and Summative Evaluation.* New York: McGraw-Hill, 1971, Chapters 1, 4, 5, 6. Comprehensive discussion of formative evaluation, its purposes and components.

DeCecco, J. P. *The Psychology of Learning and Instruction: Educational Psychology.* Englewood Cliffs, N.J.: Prentice-Hall, 1968, Chapter 14. Extremely clear presentation of evaluation in a criterion-referenced system.

Florida Educational Research and Development Program. *There Is a New School Coming: Third Annual Report*, Tallahassee, Fla.: Florida Department of Education, 1972, Appendix G. Succinct listing of the characteristics of a well-constructed criterion-referenced test item.

Frederick, N. A., Gamelo, P. S., and Gleason, S. S. *Utilization of a Catalog of Reading Objectives: Evaluation.* Chipley, Fla.: Panhandle Area Educational Cooperative, 1971, pp. 125–194. Performance-based teacher training module that follows approximately the same systems approach model as our text.

Kerlinger, F. N. *Foundations of Behavioral Research.* New York: Holt, Rinehart and Winston, 1973. Excellent resource document. Comprehensive description of different types of interviews and questionnaires and their construction, application, and interpretation.

Physical Education

Clarke, H. Harrison. *Application of Measurement to Health and Physical Education.* Englewood Cliffs, N.J.: Prentice-Hall, 1967; Mathews, Donald K. *Measurement in Physical Education.* Philadelphia: Saunders, 1968; Scott, M. Gladys, and French, Esther. *Measurement and Evaluation in Physical Education.* Dubuque, Iowa: Brown, 1959. Widely used texts covering various motor performance, physical characteristics, and written tests in physical education.

8

Learning Conditions and Applications

This chapter describes the nature of a strategy and its role in directing the student toward a specific objective based on his present status. Strategies reflect consideration for a host of factors, such as learning events, sequencing, and modes of event presentation. Although these terms and concepts may be unfamiliar to physical educators, they are gaining increased acceptance by educators in many other fields.

The teacher is faced with the difficult task of analyzing the learning situation to determine appropriate instructional strategies. He cannot plan effectively without knowing how learning conditions affect the acquisition of behaviors. Thus the heart of the chapter is divided into two sections. The first describes the GAMES plan, which graphically illustrates the primary steps in preparing the student for learning: Goal orientation, Attention, Motivation, Experienced for transfer, and Situation familiarity. The second section deals with actual learning conditions, outlined in the PRACTICE plan: Practice, Reinforcement, Administration, Contiguity, Tradeoffs, Information feedback, Cue generalization and discrimination, and Event repetition.

With the information presented in this chapter, you should be able to design effective instructional strategies for any specified terminal behaviors in the physical education class.

Student Objectives

The student who understands the material in this chapter should, with the book closed, be able to:

1. Define instructional strategy
2. Specify the four event variables in the development of instructional strategies
3. Give three examples of internal conditions of the learner with regard to learning events
4. State at least six of the nine external events usually associated with instruction
5. Define what is meant by the presentation form of an event
6. Define two kinds of event-sequencing considerations
7. Describe the relationship of learning to practice and change
8. Distinguish performance from learning, referring to temporary states of the learner
9. Describe three kinds of behaviors exhibited by a person which may interact with learning but can be considered as separate behavioral manifestations
10. Give two illustrations of limitations of learning implications
11. Identify the five factors in the GAMES plan as they apply to the preparatory state of the learner, and briefly describe the function of each
12. Identify the eight factors in the PRACTICE plan as they apply to learning conditions during practice or experience, and briefly describe the function of each
13. Describe four conditions that should be identified and manipulated in order to facilitate retention

The Nature of a Strategy

Having discussed the importance of identifying and defining instructional objectives, we must now concern ourselves with designing an instructional strategy which will enable us to achieve them. Figure 8-1 shows the location of instructional sequence in the systems approach model.

In this chapter we will synthesize recent research in education and the psychology of learning and apply it to the development of instructional sequences with specific objectives. In formulating the content of any class, you must consider these four variables:

FIG. 8-1 *Systems approach model*

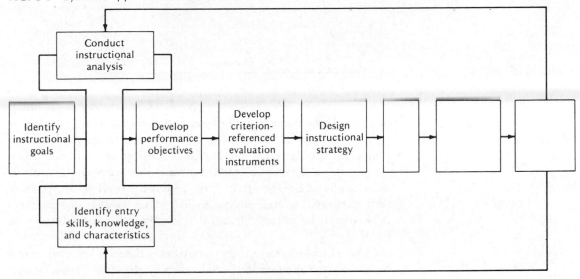

1. type of events
2. sequence of events
3. scope of events
4. presentation of events

If we interpret an event as what a psychologist might call a stimulus condition, several questions arise: What kinds of activities does a student require in order to progress effectively? How many events are necessary to insure learning? What should their magnitude and relationship with each other be? How should these events be arranged and timed? In what form should they be presented?

Instruction is actually the control of external (to the student) events by the teacher or his use of media (books, videotape, films, etc.). Since the teacher is not the only source of information, it is his responsibility to make wise and appropriate decisions concerning instructional objectives and strategies. As Figure 8-2 indicates, he must first assess student entry behaviors and characteristics relevant to the learning tasks, and then state his objectives clearly. Both these procedures directly influence his selection of instructional content, i.e., the type, scope, sequence, and presentation of events, and the formulation of his strategy. Finally, he must evaluate the strategy to determine its effectiveness.

A simple example in a physical education class may help at this point. Let's assume that you are instructing a class in diving and

FIG. 8-2 *Factors in choosing and applying strategy*

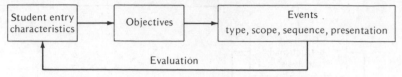

that one of the instructional objectives is executing a front somer-
sault dive from the low diving board according to performance
standards specified in the National Collegiate Athletic Association rules. First you must consider the students' characteristics as they begin to attain this objective. To what degree have
they mastered related diving skills? What is the developmental status of the group? Should it find this task relatively easy or
hard?

Once entry characteristics have been determined, you must
decide on the content and medium of instruction (type, scope,
sequence, and presentation of events). Some questions to consider
might be: Which kinds of related, enabling skills should students
master before they begin practicing the actual task? How much
time will be needed for practice on any related skill, subtask, or
instructional procedure? In what order will the related skills or
instructional procedures be presented?

Formulating a strategy with set purposes is analogous, at least
in part, to what some educational technologists refer to as designing an instructional system. (Another common term for it
is instructional process, and perhaps even teaching methodology.)
Bear in mind that there is no one strategy for all tasks. Certainly,
environmental and resource restrictions may dictate the use of
a particular strategy. Be aware of alternatives, and design strategies carefully so that students will achieve the required instructional objectives.

So many types of objectives are possible that strategies can
become complex and detailed. For example, it is one thing to
state objectives for skill improvement, movement awareness,
organic development, social interactions, intellectual growth, and
the like, and another thing to decide on the strategies for achieving
them. Strategy decisions, though not simple, are definitely easier
when objectives are limited in scope and number. This does not
necessarily mean that you should state few objectives when teaching. However, you should realize that the entire instructional
process becomes more difficult when you specify a variety of
objectives. Be idealistic but practical.

Focus

The design of an instructional system requires the use of instructional strategies. You decide what to teach, and in what order, in advance of the class and course meetings. Paul A. Twelker has contributed a chapter entitled "Designing Instructional Systems" to the National Research Training Manual. *In his introduction he expresses the importance of instructional design:*

> When we talk about the design of instructional systems, we are thinking in terms of the tasks involved in specifying in a systematic fashion a series of learning experiences that will produce consistently and predictably a desired or stated behavior on the part of the learner when implemented. A parallel may be drawn between an architect and an instructional system designer. The architect specifies guidelines and plans for each step of the construction of a building—the end product is a set of blueprints that when translated by a contractor results in a building. The instructional system designer is in one sense an architect. He specifies various components of instruction: media, content, instructional strategies, and so forth. The result is a blueprint for instruction—something that an instructional system developer might use to build a prototype from the specifications. Needless to say, the specifications should be in sufficient detail so that a person other than the designer could take them and develop the instructional system.
>
> How does one go about the complex job of designing an instructional system? Basically, two steps are involved. Given a clear statement of terminal objectives, the designer must determine the sequence of instruction. That is, he must determine what enabling objectives or *en route* objectives are required and in what order they should be taught. Then he must specify the instructional conditions that best "fit" the objectives. (1969, p. II-1)

The Learning Event

According to Robert M. Gagné (*The Conditions of Learning*, 1970), the success of a learning event in educational research and the systems approach depends on both internal and external conditions. Internal conditions determine the student's readiness and willingness to learn and are therefore prerequisites to learning. External events are the environmental conditions manipulated by the teacher.

Gagné identifies three factors which determine internal conditions: attentional sets, motivation, and developmental readiness.

Attentional sets refer to the student's ability to discriminate among available stimuli and to select the appropriate ones at the right time, an ability which determines sequences of action. As a further consideration, Gagné states that "the gaining and maintenance of attention, and the lengthening of what the first-grade teacher calls the span of attention, are instructional questions that concern the presence or absence of an attentional set." (p. 278)

The importance of motivation in learning has been acknowledged for many years. A desire to learn, coupled with an interest in the activity, increases the student's chances of success. Many factors can affect motivation. When persistent, it may be attributed to a general drive (perhaps innate) for competence, a conditioned response to experience with a variety of matter, or the reinforcing nature of achievement, self-fulfillment, and ego-gratification.

Developmental readiness depends on the student's age and his experience with mastery of other, related skills which serve as building blocks for the present tasks. If motivational limitations imposed on the student hamper his progress, the teacher must obviously give the situation serious consideration.

External events, according to Gagné, are what most people associate with instruction. He suggests that they be sequenced as follows:

1. Gain and control the student's attention
2. Tell him what is expected of him
3. Inform him of previously acquired skills and knowledge that are prerequisites for the current learning
4. Present the stimuli involved in the learning task
5. Guide and direct the student
6. Provide him with feedback to inform him how he is doing
7. Give him opportunities to appraise his performance
8. Establish the generalizability (transferability) of the newly acquired capability
9. Insure its retention

Gagné does not mention how each event might be administered, so in a sense we have an incomplete strategy. The total instructional strategy for a task should include not only events but also the degree of emphasis on any one event and its mode of presentation. Not all authorities would agree on the particular sequence of events or on the need for each event in all learning situations, but Gagné's list is general and relevant enough to stimulate your thought processes.

If we were to apply the list to a physical education activity, the following actions might be decided upon:

Objective: Students will be able to perform the crawl stroke according to Red Cross standards

Events: Prepare the students to learn the swimming task:
1. Gaining their attention and establishing motivation
2. Inform the students of the objective and the standards against which they will be appraised, and describe the way they should be able to swim the stroke after instruction has been completed
3. Make sure the students are aware of those swimming skills they already possess that will enable them to learn the crawl relatively easily
4. The stimulus in this case is the water (in the pool, probably), in which and to which students will perform
5. Instruct and guide the students in the execution of the crawl
6. Provide the students with feedback (appraisal of performance) so they know how they are progressing
7. Encourage the students to appraise their own performances
8. As they master the crawl, show the students how they can transfer (generalize) from the present swimming stroke to others
9. Finally, the extent and quality of the practice should enable the students to remember how to perform the crawl in the future

Although we may make some changes in these steps later, they exemplify adequately what we mean by instructional events.

The Scope of Learning Events

The scope of a learning event refers to the length of time a student is exposed to it. Some events contribute more than others to certain kinds of objectives. Thus the importance of a particular event determines its scope.

The number of events necessary to insure the realization of an objective should be considered. Some objectives require more events than others. When difficult skills and a high level of performance are expected, special scope considerations will be required. Special considerations are also needed for students with poor skills and low objectives, interest, intelligence, and so on.

Presenting a
Learning Event

Once you have decided on the type and scope of events, you must determine the manner, or mode, of presentation. (Instructional technologists would probably refer to the presentation mode as the *medium*, i.e., the means by which something is conveyed or effected. Any learning environment developed as such thus becomes a medium.) The presentation mode could be a lecture by the teacher. It might be self-directed, with suggested adjunct materials. Or it could be media (technological and other instructional aids) used to enhance the learning situation. (Chapter Nine deals solely with descriptions and functions of media in instructional settings.) Possibilities are endless for any situation.

The presentation mode is a selected medium (based on an analysis of the situation, available resources, and established evidence of learning) implemented in a particular way (introduced as an operational system). After a particular medium, such as videotape, is selected, it must be drawn into the instructional operation; only then will it enhance learning. The mere presence of the videotape does not insure learning.

Although presentation modes are diversified, for the sake of convenience they could be grouped in four categories. In *Educational Media and the Teacher* (1970), John Haney and Eldon Ullmer describe four types of learning activities, which we would call presentation modes: exposition, participation, interaction, and evaluation. Each activity has strengths and weaknesses, and any one might be more effective than the others as an enabling objective for a particular terminal instructional objective.

Exposition Just as the word suggests, the student is exposed to a learning experience. He is not usually actively involved in it but instead is receiving (seeing, hearing, perceiving) and processing information. Examples of this type of learning activity are listening to lectures, reading, viewing or listening to media, and the like.

Participation The student is required to perform in some way. He may be asked to make an oral or written report, complete a programmed text, execute a clear in badminton, run an experiment, solve a problem, or swim the length of a pool.

Interaction Students and teacher are involved in a dialogue. They respond to each other's ideas and feelings in a give-and-take atmosphere. Examples would be seminars, discussions, conferences, specially formed interest groups, and simulated games.

Evaluation The students and the teacher determine the quality of the educational outcomes. Assessments of activities—formulations, applications, or both—can be learning experiences in themselves.

Presentation modes for any objective may be introduced singly or in combination. Many factors will influence your decision. For example, teaching a volleyball skill might involve exposition, participation, interaction, and evaluation, or it might involve only exposition and participation. You must decide how much time to allot for any learning activity. We do not attempt to make specific recommendations here, but these four categories can serve as a guide as you attempt to formulate enabling objectives in your particular situation.

It would be impossible to list in this book all the possible presentation modes for all the events. For some events perhaps only one mode is really possible. For others, a number of modes may be possible, with one acknowledged as the best. For still others, a number of modes may be equally effective. We may also ask whether one presentation mode is as effective as the simultaneous or consecutive use of a number of modes.

For example, one of the events suggested by Gagné was to inform the student of what is expected of him. How would a swimming instructor do this in teaching the crawl stroke? He could perform the stroke himself and *show* the students what it looks like. He could ask a student who knew how to perform the stroke to demonstrate it. He might show *films* of the correct execution of the stroke. The students could *read* about it and examine illustrations of it. The instructor could *tell* the students about the stroke. Realizing these alternative possibilities, the instructor would select one or more presentation modes. His choice would usually be based on personal experience and preferences, although it would make more sense to base it on research and the availability of desirable instructional media.

Another event cited by Gagné was to guide and direct the student. Assuming that our objective still deals with mastery of the crawl stroke, list the possible alternatives for reaching this objective using that event.

_____

_____

_____

_____

Some possibilities are listed for you at the bottom of the page, but don't peek until you have attempted to respond. Now, is one way better than the rest, or would you use a combination of presentation modes? Would you emphasize a particular mode in one or more events?

So far, we have referred solely to an objective dealing with learning and skill acquisition. It would not make too much difference, however, if we had specified other kinds of objectives. Consider one in the affective domain that could be applied to the swimming class:

Objective: Students will demonstrate an increased interest in swimming by voluntarily swimming more during their free time as the class progresses.
Events: Many of those already stated are quite acceptable.
Presentations: A number of possibilities.

Let us look at a particular event, to guide and direct the student, and examine some of these alternatives. For the present objective, you might tell students when the pool is open for recreational swimming. You might encourage them to swim then by saying that the class time is not adequate for the practice they need to master the activity. The students should have good experiences (minimum danger and accidents, conscientious teaching, warm teacher-student relationships), which will positively affect their attitudes toward and interest in swimming. And doing so on their own time.

Students will not automatically develop positive attitudes, and some may even change their attitudes from positive to negative during the course of instruction. So if one of the teacher's objectives is to develop more favorable attitudes toward swimming, and one of the criteria for evaluating this objective is time spent in recreational swimming, then class situations and strategies must be developed that will ensure its realization.

The important point to remember is that the specified instructional objective will encourage the use of certain strategies: that is, a particular sequence of events and their modes of presentation.

Verbal instruction from instructor
Mutual assistance by students
Movement of passive student's limbs by instructor
Films
Flotation devices and various other aids

Sequencing Learning Events

We would be remiss in leaving a discussion of instructional events and sequencing without explaining two kinds of sequencing maneuvers. In this chapter sequencing has been related to events. We stated that one of the teacher's responsibilities is to order and arrange the sequence of events which the students will experience.

In the previous case, sequence of events referred to environmental or practice conditions manipulated and controlled by the teacher. However, sequencing can also describe a hierarchy of tasks which might be ordered for maximal positive-transfer effects. For every end-of-class, or terminal, objective, there are enabling objectives that must be attained if learning is to progress rapidly and orderly. A series of experiences is designed for each enabling objective, and in turn the enabling objectives lead to the successful realization of the terminal objective. The learning of one competency assumedly influences positively the learning of another.

Not all learning contains dependent and interrelated elements. In such cases instruction may very well proceed in an arbitrary manner as the sequence or order of presentation will make little difference. Sometimes it is difficult to ascertain whether tasks are independent or interdependent. Leslie Briggs, in *Sequencing of Instruction in Relation to Hierarchies of Competence* (1968), suggests that the answer would be task analyses in the class of interest. There is a need to discover the optimal sequencing of instruction, for research in this area is lacking. Although it is generally agreed that sequencing of instruction is important in the case of dependent units, "views differ widely as to *why* sequencing is important, and also as to how to provide effective sequences of instruction." (Briggs, 1968, p. 26)

Recall, if you will, our objective of having the students perform the crawl stroke according to Red Cross standards. That is a terminal objective. Enabling objectives (which help the student master the terminal objective) might be (a) submerging the head, (b) gliding in the prone position, (c) doing the flutter kick in the prone position across the width of the pool, (d) etc., etc., etc. A suitable sequence of tasks is arranged to encourage learning and positive transfer. Thus you can see that the sequencing of tasks and the sequencing of instructional procedures are distinct, yet compatible, responsibilities.

Instructional Variables

When we talk about objectives and strategies for reaching them, the intervening variable involved is learning designated skills, knowledges, and attitudes. Learning has been defined in many ways, and a fairly general but usual interpretation has been ex-

"The learning of one competency assumedly influences positively the learning of another."

pressed by Singer in *Motor Learning and Human Performance* (1968): "the relatively permanent change in performance or behavioral potential resulting from practice or past experience in the situation." (p. 3) Change and practice or experience are thus factors in the learning process. Therefore:

practice or experience $\longrightarrow$ learning $\longrightarrow$ relatively permanent change in performance or behavioral potential.

Learning Considerations

The purpose of this chapter is to help you facilitate learning by insuring that practice and experiences take place under the most favorable conditions. Objectives concerned with athletic skills and movement patterns, social behaviors, attitudes, intellectual content, physical fitness, and the like are met when the students' characteristics and the learning process are given equal consideration. Whether the behaviors rest primarily in the psychomotor, social, cognitive, or affective domain, learning factors are fairly consistent, and a general framework can be developed to show them. It is also important to identify any unique factors related to the classification of a particular objective in a particular behavioral domain.

Since this book deals with teaching and changes in behavior and performance, it must be based to a great extent on research dealing with the nature of learning and the enhancement of the learning process. Students differ widely in prior experiences and degree of competence, age, sex, body build, physical characteristics, motor abilities, personality variables, and levels of expectation. These and a host of other variables interact with the learning environment established by the teacher.

Although we can and will identify learning phenomena generally considered extremely relevant to changes in behavior and performance (the result of learning), we should emphasize the fact that not all students, because of the nature of individual differences, respond to and benefit similarly from the same application of learning factors (as represented by one instructional strategy). Nevertheless, factors that affect the learning process are fairly predictable for most of us. They are translated into learning events and specific instructional decisions. The more effective teacher identifies the salient learning factors, sequences them in the form of learning events, and probably offers a variety of presentation modes for a number of events. Students can thereby select instructional media compatible with their personal styles of learning. This procedure insures that all students will benefit from their experience with key learning features.

Figure 8-3 identifies some of the key elements you should consider to insure learning in every situation. The more homogeneous the group of students with regard to entry behaviors and characteristics, the less concern need be shown for individual differences, and the more successful a particular instructional strategy might be. On the other hand, a heterogeneous mix will benefit more from a different kind of instructional strategy. For instance, the more highly skilled can assist the less skilled. Figure 8-3 also shows a number of the learning variables which must be sequentially placed into events and then incorporated within the practice sessions as instructional strategies. The ideal intermix of instructional strategies with learner characteristics produces ideal changes in performance and behavioral potential.

You should always keep in mind that practice by itself does not insure learning. If the student does not know what he is doing, why he is doing it, or how well he is doing it, why should he spend time practicing? Nondirected, unmotivated, and meaningless experiences, no matter how much time is allotted to them, are sure to produce meaningless results. Positive changes do not occur

FIG. 8-3 *The intermix of learner status, practice, and strategy variables in the learning process*

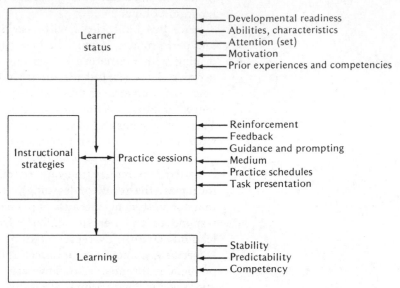

in skill development, for example, unless the learning situation is arranged in a satisfactory manner.

Furthermore, we all realize that at first, the performance of skills is rarely stable and predictable. As we become more skilled, stability and predictability both increase. In the earlier definition of learning (see p. 174), reference was made to the relative permanency of performance or behavioral potential. The more easily performance (tests of motor skill, knowledge) is disrupted, the more probable it is that learning levels are not too high. If behavioral potential—the tendency to react to a given situation in a certain way—is not strengthened, it too will be easily disrupted by changing circumstances. Tendencies to respond, measured indirectly, are usually conceived of as attitudes and values.

Performance Variables   We have identified a number of performance variables, so called because they temporarily elevate or depress performance. Consider the conditions shown in Figure 8-4. Any one of them can cause transient fluctuations in behavior and thereby conceal true learning levels. A student who is sick, sleepy, nonconcerned, or hurt will not perform to the extent of his capabilities. The lesson here is that performance does not always truly represent learning status. But with all its limitations, performance is the best and most often used measure of learning. If true learning levels are to be determined, try to make sure that performance variables are at a minimum and student motivation is optimal.

FIG. 8-4 *Performance variables that contribute to temporary states of the learner*

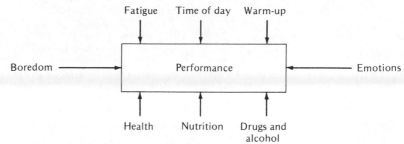

Since we are attempting to describe the nature of learning and performance, their interaction, behavioral potential, and learning and performance factors, we should point out that some behavioral expressions are not examples of learning. One category has just been discussed: *temporary states* of the organism.

Another category of behaviors could be related to *maturation* and growth and development factors. A child cannot learn complex behaviors until he has reached the appropriate stage of neurological development and maturation. Certain behaviors tend to be evinced as the child develops. Most are probably the result of the interaction of learning and maturation. Teachers concerned with change and learning, however, should be able to attribute behavior modifications to instructional processes rather than simply to maturation. It might be interesting to speculate about performance changes in younger students after a year of physical education classes. Have they changed because of the instruction or simply as a function of maturation and in spite of the instruction?

The third category of behaviors may be classified as *reflexes*. In every species unique reflexes are genetically transmitted. The knee-jerk reflex, like many others, is not learned. Reflexes, maturation, temporary states, and learning can each produce relatively independent and unique behaviors. Interactions among them are very possible, as revealed in Figure 8-5. Performance and behavioral potential at any given time can reflect the incorporation and interaction of these factors.

Learning Conditions and Applications

Conditions affecting learning have been the subject of many books on the psychology of learning and the psychology of teaching. The research literature is replete with articles dealing with aspects of learning in a variety of situations and with numerous tasks. Research findings and the leading learning theories give us valuable insights into the nature of the learning process and suggest how to teach effectively.

FIG. 8-5 *Behavioral expressions which may interact with learning*

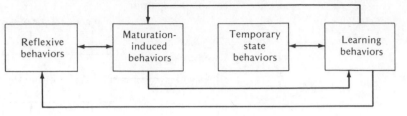

Limitations

But a word of caution is needed lest we become overly confident in the answers researchers have provided for us. Instructional strategies can no doubt be improved considerably by an understanding of the various aspects of learning. However, we still need more direction in effective teaching and subsequent learning.

Focus

We do not mean to imply that research and theory are meaningless, that we can teach the way we want, using only intuition, common sense, and past experiences. In The Conditions of Learning *(1970) Robert Gagné intended to inform his readers about what is known about the effect of the learning process on instructional outcomes. The association of learning and instruction is apparent throughout the book. He also presents a section on "Limitations of Learning Implications":*

The reader needs to be made aware that there are some problems of great importance to education which *cannot* be solved by applying a knowledge of the principles of learning as they are described. For example, there are many aspects of the personal interaction between a teacher and his students that do not pertain, in a strict sense, to the acquisition of skills and knowledges that typically form the content of a curriculum. These varieties of interaction include those of motivating, persuading, and the establishment of attitudes and values. The development of such human dispositions as these is of tremendous importance to education as a system of modern society. In the most comprehensive sense of the word "learning," motivations and attitudes must surely be considered to be learned. But the present treatment does not attempt to deal with such learnings, except in a tangential sense. Its scope is restricted to what may be termed the intellectual, or subject matter content that leads to improvement in human performances having ultimate usefulness in the pursuit of the individual's vocation or profession.

From *The Conditions of Learning,* second edition, by Robert M. Gagné. Copyright © 1965, 1970 by Holt, Rinehart and Winston, Inc. Reprinted by permission of Holt, Rinehart and Winston, Inc.

Another kind of limitation needs to be mentioned. Regardless of how much may be known about how to *begin* the process of establishing competence through learning, it is clear that no one knows very much at present about how to *continue* the process to its highest levels. It does not seem possible at present to specify all the conditions necessary to attain the highest and most complex varieties of human performance such as those displayed in invention or esthetic creativity. How does one produce an Albert Einstein or a Leonardo da Vinci? Certainly there are distinct limits to currently available knowledge bearing on such questions. The most that can be said here is that the production of genius is not based on "tricks," but on the learning of a great variety of specific capabilities.

To understand how learning operates in everyday situations of the school is a most valuable kind of understanding. But it does not unlock all the mysteries of education. What it can do is to illuminate some of the activities of the curriculum planner, the course designer, and the instructor. (pp. 25–26)

Key Conditions

Limitations realized, let us turn to a variety of learning considerations, for the conditions that can be manipulated by the physical education teacher or coach are unbelievably extensive. We will mention only those conditions we deem most worthwhile.

The format we have decided to follow for this presentation is generally one that sequences learning conditions in the order in which the teacher might sequence learning events. Each learning condition may not equally affect the realization of objectives in the different categories of behavior—cognitive, psychomotor, social, and affective. Some special considerations may have to be made for objectives within any category.

We should also point out that learning of some kind occurs regardless of the method used—problem solving, programmed texts, or trial and error (fairly student-centered methods); or drill, preferred prescribed steps and strategies, or generally traditional behavioristic (teacher-centered) methods. The way an objective is stated could dictate the use of one of the preceding approaches, or perhaps another. It is assumed that in choosing a teaching method the teacher will incorporate his knowledge of learning conditions that will favor the achievement of certain objectives. One must always weigh costs and payoffs in any instructional decision.

Should the emphasis be on the final product or the process? If acquiring the techniques involved in a learning process—e.g., problem solving—is important, the product, in the form of specific behaviors, becomes secondary to the process, which will be applicable to other situations. Once again, consider objectives carefully, for they help to dictate strategies, including teaching methods.

Although the systems approach is not meant to confine the

teacher to one teaching method or strategy, the learning conditions and instructional events to be presented will reveal a trend toward a prescriptive approach along behavioristic lines. We feel that most teachers want examples of specific procedures to follow when they attempt to affect the learning process, and we have presented and used such examples in a manner reasonably consistent with research and theory. The impossibility of discussing every possible alternative must be obvious. However, you will obtain the best results if you view instruction in a systematic way, with concern for process and product as well as the other factors emphasized in this section of the book.

Preparatory State of the Learner: The GAMES Plan

Although many preexperience considerations can and probably should be made, the teacher must determine priorities due to time and other limitations. We have attempted to identify five preexperience learning states that are important in learning almost anything. These factors are incorporated in what we call the GAMES plan. For best possible results the learner should be:

Goal oriented
Attentive
Motivated
Experienced for transfer effects
Situation acquainted

Goal Orientation   We know that children learn by modeling their behaviors on those of adults and other children. The teacher can inform the students what is expected of them by means of observation–demonstration techniques, which provide them with an understanding of where they are headed. There are various ways of directing students' goals, including the use of written material, films, illustrations, verbal descriptions, and observation.

No matter what means are used, the student can be goal-oriented only if he understands the goals. He must create an image for himself of the desired response. How does the teacher hope the oncoming learning experiences will affect particular behaviors or tendencies to behave? Where instructional strategies such as problem solving are advocated and students are led into the act of discovery, goals and objectives need not be specified. The choice of strategy depends to a great extent on what the teacher is attempting to achieve. Specific skill expectancies require specific explanations. If a number of behaviors of a given type are acceptable, then goal direction need not be so explicit.

No matter whether our objectives in changing behaviors are related to cognition, motor and physical development, social relations, or attitudes, we should provide students with a reasonable degree of goal orientation. The nature and extent of this orientation will depend on the particular behaviors to be modified as well as on available sources and resources.

A popular technique in physical education is imitation, in which the student attempts to copy the behavior or performance of a model (the teacher, an expert student, or a person in a film). Psychologist Albert Bandura has extensively researched the topic of behavior modeling, indicating how behaviors are acquired and modified as a consequence of this process. The terms modeling, copying, imitating, and observing are used interchangeably. In *Principles of Behavior Modification* (1969) Bandura makes the following sweeping statements about the effects of observing another person's behavior:

> . . . one can acquire intricate response patterns merely by observing the performances of appropriate models; emotional responses can be conditioned observationally by witnessing the affective reactions of others undergoing painful or pleasurable experiences; fearful and avoidant behavior can be extinguished vicariously through observations of modeled approach behavior toward feared objects without any adverse consequences occurring to the performer; inhibitions can be induced by witnessing the behavior of others punished; and, finally, the expression of well-learned responses can be enhanced and socially regulated through the actions of influential models. (p. 118)

Attention   It may not always be necessary for the teacher to use special means of directing students' attention to the behaviors to be learned, especially with more experienced and mature students. But with younger, inexperienced students, the teacher must usually provide a condition that gains and maintains their attention.

Attention implies interest and motivation as well as the ability to discriminate among available stimuli and select those to which a response is expected. When students are taught to hit a softball, they have to turn their attention consciously to the task through their own efforts or as directed by the teacher. High attention levels are important during all stages of the experience, but especially when students are first introduced to the task. One of the prerequisite conditions for learning is that the student be in a state of preparedness, alertness, and attentiveness. Receptivity to directions and concepts can be translated into effective behaviors. Various techniques have been suggested to direct the student's attention to a task if he is not demonstrating it on his own initiative. Whatever the strategy selected, once the student is attentive, he is more apt to acquire the specified behaviors.

Closely related to this discussion of an external (to the learner) learning condition is the internal condition Robert Gagné (1970) calls attentional set (see p. 167), which precedes the learning event itself and which denotes a state of readiness for learning. Although it is an internal condition, the attentional set can be externally managed and directed by external agents, such as the teacher. Gagné writes that the student

> . . . must learn to continue to direct his sense organs toward the sources of stimulation, to discriminate the essential features of that stimulation, to maintain the internal cues that determine sequence of action, and to continue an activity that he has begun in the face of distractions, some of which may come from his own body. (p. 279)

These capabilities, according to Gagné, can probably be classified as attitudinal sets and may be considered essential to future learning of similar matter or in similar conditions.

Motivation Any consideration of the preparatory state of the student must include his level of motivation. The sensitive and knowledgeable teacher realizes that the student performs most readily and skillfully when ideally motivated. Motivation can be influenced by many variables and in turn can affect the learning and performance of a variety of tasks.

When a person is activated toward some goal, he attempts to reduce the drive by achieving it. Thus motivation prevails until he accomplishes the goal or no longer deems it essential. How does a teacher encourage the student to want to demonstrate some behavior, and then keep him motivated? Sometimes the learning situation itself contains enough challenge, satisfaction, and rewards to stimulate the student in the desirable direction. Children do not have to be forced to play. They have a natural tendency to express themselves through psychomotor behaviors. In this respect the physical education teacher working with young children faces a most enjoyable task. He would have to work hard to destroy a child's interest in and motivation to play.

Yet for some reason, many children's favorable attitudes toward physical education and physical activity seem to diminish as they grow older. This unfortunate circumstance can be due to a number of factors. Whenever physical education is required instead of elected, as with older students, motivation will probably be more of a problem to the instructor. Although it is true that the student comes to the learning situation with a preestablished, internally set level of motivation, it is equally true that the teacher can directly or indirectly influence this level in the instructional situation.

Apparently, a student can be either intrinsically or extrinsically motivated, or both (Figure 8-6). Educators have usually tended to favor the former method, in the belief that a student will be more highly motivated to accomplish a task if it will give him fulfillment, happiness, and personal satisfaction. Extrinsic incentives, such as prizes, rewards, grades, and the like, have also been shown to be effective modifiers of behavior. It is to be hoped, however, that with practice and experience students will learn to depend less on external prompting and more on self-direction. Apparently, the learning of skills, attitudes, and information in physical education can be influenced by a variety of circumstances, and the effective teacher realizes and uses those motivational techniques appropriate for the class as a whole as well as for individual students.

If the student is demonstrating an ideal level of motivation as he enters the physical education class, the problem is to maintain it. A favorable learning situation that leads to achievement and satisfaction will increase the likelihood of sustained motivation. The entering student who demonstrates a poor attitude toward physical education will require a different approach.

Stating class goals, and rationales for them, may increase motivation. Older students especially want to understand *why* certain behaviors are expected of them, to make sense of learning situations.

Tasks should be set at the students' level. When a student succeeds, he becomes interested in improving his performance. But when he fails, he becomes frustrated, loses interest and motivation, and stops participating in the activity. People like to do things they can do reasonably well, especially in the presence of others. When students are in class, they are not learning and performing independently but are influenced by the teacher and other students. Less able students will need to be motivated more than others, unless for some reason they are sufficiently goal-directed as

FIG. 8-6 *The interaction of intrinsic and extrinsic motivation in human behavior*

they enter the class. In any event, it is important to consider the following factors, which influence student motivation: (1) a hierarchy of objectives, each of which can be attained after reasonable experience; (2) objectives and rationales; and (3) rewards, reinforcement, and feedback.

Also, learning research indicates that students are not equally motivated by the same incentives. What you perceived as an excellent source of motivation may not be perceived in the same way by your students. For instance, if you offered five dollars to the student who learned how to shoot the highest percentage of foul shots in basketball, wealthy students would not respond as enthusiastically as those who were less well-to-do. Grades may stimulate some and fail to interest others. Thus you must choose incentives wisely for the group as a whole and make exceptions for individuals if necessary.

One last point: there is no one best motivational level for all students and all tasks. There is, however, an ideal level for each student and each task. There is no logic in the statement "The more motivation, the better the resultant performance." Behavior is far too complex for such simple observations. Rather, consider (1) the emotional characteristics (anxiety level, stress resistance) of the student, and (2) the difficulty and complexity of the task. Then consider (3) whether motivation is needed from external sources, and if so, how much.

If a student is very anxious, additional, external motivation may be quite upsetting, especially if the task is reasonably difficult. Consider a nervous, high-strung student attempting to learn to play golf, while the teacher provides various inducements to elevate motivation. This student needs to relax before he can acquire skill. You should always analyze the nature of the behaviors to be learned in order to determine their potential difficulty. An observation appraisal of each student in the situation will then indicate whether or not you need to apply motivational techniques.

Experienced for Transfer Effects It could be argued that after the earlier years of life we never really learn anything new. Childhood experiences contain the basic building blocks for all we will learn for the rest of our lives. Through the years more fortunate individuals will encounter more varied and enriched experiences. The nature and extent of their prior, relevant experiences will affect the ease with which new learning occurs.

The term usually used to denote the relationship of prior experiences to new ones is *transfer*. For convenience it is logical to describe a task never directly practiced before as new. But in keeping with our argument, think about the possibility that what makes a task "new" is simply the rearrangement and restructuring of previously learned experiences. Perhaps every "new" task or situation can be simplified by identifying those basic components which are part of the student's existing repertoire of behaviors.

Accordingly, the nature of the relationship between previously learned and new tasks will show positive or negative transfer. In other words, present learning will be enhanced or inhibited as a result of prior experiences. In experimental situations the term zero transfer is also used. That is, no relationship exists between situations; consequently, new learning is not affected in any way. It is easier to contrive such situations with artificial tasks in a laboratory under controlled conditions than to find similar examples in real life. We might decide that there was no relationship between completely different everyday tasks. But an appraisal of the dynamic interplay of *all* prior experiences of the organism will probably indicate an expected generally positive or negative effect on the new task. So we will consider only beneficial or detrimental relationships between situations.

The "new" task may require previously learned skills, but in greater complexity, altered temporal and sequential patterning, and in a different environmental context. The degree to which positive or negative transfer will occur between tasks will be determined by a host of factors, among the most important of which are the similarities between the stimuli components and the response components. Without becoming technical, we might mention that positive transfer is quite powerful and sustaining, while negative transfer is fairly temporary and transient. This is indeed fortunate.

Think about any two activities and attempt to predict transfer effects. For example, if a person had played a lot of tennis in the past and was about to learn badminton, what kind of transfer would you expect? Why?

_____

_____

_____

_____

_____

We would probably expect a slight overall positive effect. Both sports involve movements, tactics, and somewhat similar strategies; and both are court sports, requiring implements and projectiles. In these ways, prior tennis experience would benefit the acquisition of badminton skills. On the other hand, the shots are technically hit differently, the projectiles require a unique touch when contacted, and differences in court dimensions and play demand altered behaviors. Thus transfer effects would be both positive and negative, although probably more beneficial than limiting.

What about relationships between baseball and softball? The effects may be somewhat similar. Think of other sport-to-sport, skill-to-skill relationships. Think also of the transfer of generalized movement patterns to activities demanding precise skills.

It is one thing to have experienced prerequisite movement patterns and skills and another *to see the relationship* between them and the new situation. Understanding relationships and intending to transfer strongly influence present performance outcomes. Consider the throwing pattern. Children learn to throw objects as preschoolers. Do they think of this pattern when they learn in later years to serve a tennis ball, throw a football, hit an overhead volleyball serve, or execute a badminton smash? Projectiles, implements, and complex movement patterns often obscure relationships. But when the student realizes the associations, he finds newly introduced tasks much easier to master.

Situation Familiarity   Prior to actual activity, you should familiarize the student with the immediate situation and present him with the learning environment and the stimulus cues to which responses must be learned. Pertinent facilities, equipment, objects, other people, and the like are possible sources of stimuli. The familiarization process allows the student to view the stimuli and to begin to make generalizations and discriminations.

In the case of softball, the student will be expected to perform within the markings of the playing field, with bases and base paths serving as further points of demarcation. Particular softball skills will require appropriate responses to (a) pitched balls (batting), (b) hit balls (running or catching, depending on whether the student is the batter or fielder), (c) caught balls (throwing), (d) officiating calls made (accepting or possibly questioning moderately), and (e) teammate error or team loss (accepting). The student learns behaviors when he can generalize from the conditions and sources of stimuli with relationships understood. With more advanced levels of expected behavior, the student must be able to detect stimuli and discriminate easily among them in order to respond appropriately.

The GAMES plan is a prerequisite for the learning conditions that follow. The student must be prepared for his learning experiences, as any subsequent experience is apt to be more meaningful if he is Goal oriented, Attentive, Motivated, Experienced for transfer effects, and Situation acquainted (GAMES).

". . . (badminton) shots are technically hit differently . . . and differences in court dimensions and play demand altered behaviors."

Practice or Experience Considerations: The PRACTICE Plan

The teacher can control a variety of practice conditions, with the result that appropriate learning experiences will lead to desired ends. Eight major factors are presented here. The list is not exhaustive, but it certainly represents the more important instructional considerations that will ultimately influence student behaviors. Expressed in the form of the PRACTICE plan, they are:

Practice—determining quality and extent
Reinforcement—shaping desired behavior
Administration—distributing task and class practice schedules
Contiguity—sequencing stimulus-response units
Tradeoffs—emphasizing type of practice
Information feedback—conveying knowledge of performance results
Cue generalization and discrimination—clarifying the nature of stimuli
Event repetition—choosing between specific and variational

Practice or Experience   A necessary and prerequisite condition for the demonstration of behaviors is practice or experience. Extensive, high-quality practice is necessary in acquiring and perfecting skills.

Have you ever heard of an outstanding athlete who has not spent a considerable amount of time conscientiously practicing?

The more difficult cognitive and psychomotor behaviors require more time and ideal conditions for practice. When tasks are easier, performance expectation levels are lower, and there is minimal concern for long-term retention. A certain number of cue-response associations are necessary in some form, however, if learning or behavioral potential is to be realized to any extent. With cognitive behaviors, learning usually proceeds in a non-self-testing situation. Psychomotor skills, on the other hand, are practiced, and the student can evaluate his performances regularly. So the typical learning of athletic skills is quite informative to the student, while the learning of cognitive material is not (except during formal tests). The development of attitudes and the learning of social behaviors are usually more challenging than other kinds of behaviors for the teacher to shape. They are not practiced in the same way skills are, and behavioral potential must often be inferred because it is difficult to measure precisely. Nevertheless, attitudinal and social behaviors are rarely modified in class unless the teacher plans student experiences.

High-quality practice or experience is a prerequisite for realizing any objective, be it primarily psychomotor, affective, social, or cognitive. The actual learning conditions, events, and strategies bear some relationship to the teaching of all kinds of behaviors, although a particular behavior may have its precise requirements. Physical educators and coaches often use the expression, "Practice makes perfect." Actually, this statement is only partially true. It should read, "Good practice leads to perfection." The desired achievement will be realized within the capabilities of the student and as a result of a learning environment conducive to producing positive changes in behavior.

The learning conditions discussed on the following pages are primarily associated with the acquisition of psychomotor skills and traditional forms of cognitive material. Repetitious practice and other adjuncts are necessary for these behaviors but not for others, which we will examine later.

Reinforcement   Why do we persist in some behaviors and abandon others? Our actions produce either positive or negative reinforcement. Positive reinforcement tends to increase the likelihood that the action, or at least some similar action, will be repeated. Negative reinforcement is obviously used with the intent of fading out, or eliminating, undesirable responses.

Reinforcers can assume a variety of shapes, events, and conditions, praise and criticism being two of the most often used types.

In physical education classes, high grades, prizes, teacher comments, scoring a basket, hitting a ball, and the like, are potential positive reinforcers. They inform the student that he is expressing appropriate behaviors. When the student is criticized, punished, or given a poor grade, he will tend to stop his negatively reinforced behaviors.

B. F. Skinner has demonstrated the value of reinforcement in behavioral technology. Attitudes, cognition, and skills may all be affected by appropriate reinforcers. The concept of shaping behavior suggests that any actions expressed in the desired direction should be reinforced so that the student will increasingly demonstrate behaviors leading to the objective and finally the objective itself. Thus the teacher who wisely uses reinforcing techniques can influence students and their behaviors to a considerable extent. Attitudes can be slowly changed, skills and physical fitness gradually improved upon, cognitive behaviors steadily increased, and social interaction and communication processes developed as the behavioral technologist—the teacher—shapes students' experiences.

Exactly how powerful an influence reinforcement is on learning is difficult to assess. Some psychologists swear by it, while others feel that its importance is grossly exaggerated. It has yet to be proven that humans can be as effectively manipulated and channeled as animals through reinforcement contingencies. Yet reinforcers can be used to reward and motivate students, inform them of the appropriateness of their responses, and guide them to certain goals.

Physical education activities often have built-in reinforcers (a form of feedback, also called knowledge of results). The student can see if the arrow goes into the target; a shot in the 9 circle is very reinforcing, a shot in the 3 slightly reinforcing, and a complete miss negatively reinforcing. Thus you can see that reinforcement need come not only from external sources, for example, the teacher.

When dealing with reinforcers, the instructor must always consider the following variables:

1. the presence of internal and the need for external reinforcers
2. the use of appropriate kinds of reinforcement
3. schedules of reinforcement
4. individual perceptions of the reinforcement

The more responses the student himself can evaluate as reinforcing, the less he needs external reinforcers. This situation would probably arise more often with highly skilled athletes than with a beginning physical education class. And, of course, the teacher need not be the only external source of reinforcement; other students and various forms of media may also serve the purpose. As to kinds of reinforcement, there are many, both positive and negative. The type selected would depend on the circumstance and would require an understanding of the student's reactions to one kind or another. As far as schedules are concerned, reinforcement can occur regularly after each act, at preset intervals, or at random. Certainly, in a class situation great regularity is almost impossible unless you employ individual learning techniques involving the use of programmed texts or computer-assisted instruction. Finally, not all students interpret reinforcers in the same way, to the same extent, or as having the same worth. Sensitivity to individual differences will help you utilize individual techniques, although most students in any group would tend to respond in a predictable direction with the use of standard, acknowledged reinforcers.

Administration   The management or administration of class experiences, as well as class schedules, is another learning condition warranting consideration. Although the teacher must make many management decisions, we will examine only a few at this time. A number of management procedures are established for the teacher by others, such as the school administration. A good example would be the length of each class period and the number of weekly meetings. In more innovative schools, however, class scheduling is less controlled and traditional and more flexible, with allowances for teacher input.

Coaches can probably decide on meeting schedules more easily than teachers. An interesting question might be: Is it better to provide short, frequent practices over a longer period of time, or fewer but longer sessions over a shorter period of time? The teacher or coach is interested not only in providing a period long enough for the attainment of specified behaviors, but also in proportioning the time so that it will yield the best immediate and future results. The teacher obviously has a direct say in the relative amount of time spent within a class for practicing a skill, resting, or working on another behavior. He might choose *massed practice*, i.e., repeated rehearsal of a task with little if any rest, or *distributed practice*, which is spaced out over a longer period and interspersed with rest periods or other kinds of learnings. For example, students could practice only batting during a 45-minute period. Or the teacher might decide on an alternate strategy of allotting some time for batting, some for other skills, and perhaps some for rest or for cognitively or attitudinally oriented behaviors.

The literature suggests that some form of time distribution for task practice, as well as practice schedules, is favored over massed practice. Presumably such a procedure minimizes fatigue and boredom and yields the best performances. Although the research consistently bears out this suggestion when performances are analyzed soon after practice, the long-term effects are inconclusive. That is, if students who had practiced under either massed- or distributed-practice schedules stopped practice for about a week, a month, or a year, they would probably perform at similar levels when they resumed practice. Common sense suggests that task complexity, time and space limitations, students' ages, maturation and skill levels, and the dangers involved will influence practice schedules within a class period. With tasks that are not too difficult or dangerous, less time available, and more mature, motivated, and skilled students, massed experiences in some form should certainly prove no worse than distributed experiences. As to the distribution of meeting dates, apparently shorter practice periods with shorter rest periods extended over a longer period of time are most beneficial in theory, although research is lacking on the topic. The statement probably applies primarily to beginners.

Contiguity   Contiguity refers to the timing of responses to cues, and behaviors to other behaviors. To be most effective, a particular response to a particular cue should occur almost immediately. In his well-known conditioning experiments Pavlov caused dogs to acquire reflexes through the process of contiguity. The classic example is the dog which learned to salivate at the sound of a bell. One investigation will serve as the example here. Each time the dog was fed, a loud bell was rung. Pavlov was able to condition that dog to salivate at the sound of the bell even when no food appeared. This experiment is quite removed from the behaviors found in the physical education class, although certain students unfortunately show a "conditioned" response when they drop everything and leave the class when the warning bell rings!

Although you will not really be concerned with acquired reflexes, many other relevant behaviors can only be shown when contiguity occurs. For very simple acts, the appropriate response should immediately follow the stimulus. In more complex activities, the units must be logically sequenced and timed. In behavioral terms, think of contiguity as a stimulus-response (S-R) unit at its simplest level. Almost all the skills that we will be concerned with involve *chaining;* that is, sequencing S-R units so that each follows in its rightful order. When one unit is executed, it triggers the next, and so forth until the total act is completed.

When serving a tennis ball, (a) as the body rocks (S_1), the tennis racket is brought behind the head and the ball is tossed up into the air with the opposite hand (R_1); (b) the ball in the air (S_2) is responded to by contact of the racket with the ball (R_2); (c) contact with the ball (S_3) leads to a downward follow-through with the racket (R_3); (d) the follow-through (S_4) is then the stimulus for the forward motion of the body as the player goes toward the net (R_4). Each response becomes a stimulus for the next response. The completion of a unit stimulates the enactment of the next unit in the chain of behavior. A breakdown in any of the links results in incomplete and probably inappropriate behavior, thereby demonstrating the importance of contiguity in skilled performance.

That contiguity is a necessary learning condition is demonstrated in other ways as well. Whenever behavior is modified according to principles of reinforcement, the application of the reinforcer to the desirable response must be carefully timed. Reinforcement has been discussed. Suffice to say here that one of the essential factors in shaping behavior in a particular direction is the nearly simultaneous occurrence of the reinforcing agent and the acceptable response.

Contiguity is demonstrated in other areas of learning. John De Cecco summarizes the possibilities in *The Psychology of Learning and Instruction* (1968):

> Contiguity is an important condition in most learning. Classical conditioning involves contiguity of the conditioned and the unconditioned stimulus. Operant conditioning involves contiguity of the response and the reinforcing stimulus. Skill learning involves contiguity in the occurrence of the various links in the motor chain. Concept learning involves contiguity in the presentation of examples and nonexamples. Principle learning involves contiguity in the recall of the component concepts. Finally, problem solving involves contiguity in the recall of the component principles. (p. 249)

Tradeoffs Many of your managerial decisions concerning the instructional process may profoundly influence the acquisition of behaviors. Sometimes situations have tradeoff possibilities. Massed versus distributed practice is a good example; hopefully you will make the right decision regarding the learning of the selected task. Other major issues are part versus whole practice, mental versus physical practice, speed versus accuracy practice, and real versus simulated practice.

These tradeoff decisions need not be made for every kind of behavior, and others are possible for specified behaviors. However, they are good examples of the simplest levels of dichotomous strategies, from which one must be selected. Of course you may use combinations of strategies for tradeoffs at any time. Let us discuss first the part-method versus whole-method controversy.

"A breakdown in any of the links results in incomplete and probably inappropriate behavior."

Should a task be taught and learned in its entirety, or should it be fractionated for most effective results? When you teach the golf swing, you can proceed on one of two courses (although modifications and combinations are possible):

Whole method: The students practice the golf swing from beginning to end

Part method: The students first practice the grip, then the part back swing, then the rest of the back swing, then the forward swing to execution, and finally the follow-through

Which technique—whole or part—would you advocate? Why? _____

_____

_____

_____

_____

Actually, golf experts might not agree on the answer to this one. Most people would probably concur that the simpler skills should be taught with the whole method and the more complex and difficult skills with the part method. In other words, don't "break down" a task unless you have to. Do we have to for the golf swing? It is hard to say. Analyze the *complexity* and *organization* of the task before making a decision. Task organization refers to the interrelationships of the task components; task complexity, to the difficulty the student might have learning and remembering it. To determine task complexity, consider the sequence of acts and what occurs during any portion of it. In the elementary backstroke, the arms move together in identical movements, likewise the legs, while the trunk is relatively motionless. This is a relatively simple task with high organization; thus the use of the whole method seems most appropriate. In the crawl a flutter kick is synchronized with arm strokes and proper breathing techniques. Since the task is fairly complex, with a low organization, the use of the part method is indicated.

The conceptual framework for a task is:

Whole method: High organization and low difficulty
Part method: Low organization and high difficulty

If both components of a task were high or low, it could probably be taught with either approach. Task analysis would suggest the appropriate strategy in such a case.

Another consideration is how much time students should spend in deliberate mental rehearsal of a skill and how much in actual (physical) practice of it. At first glance this may not seem to be a tradeoff problem. We might be tempted to say the more physical practice the better. Yet outstanding athletes often admit that they mentally rehearse, or conceptualize, an act before and after they actually execute it. They become prepared, they analyze, and they hope to improve their performances. Learning any athletic skill involves the interaction of cognitive, affective, and psychomotor behaviors, although the student may not be aware of it.

Since research consistently reveals that motor tasks can be learned to some degree without actual performance but by means of deliberate mental rehearsals of them, overcrowding and limited equipment in a class could suggest formalized mental rehearsal sessions as an alternate strategy to sitting and doing nothing. At times brief mental rehearsal situations prior to actual performances may enhance the learning process. It does seem, however, that the true beginner might benefit less from mental practice than the more highly skilled, perhaps due to the former's inability to conceptualize the desired response correctly.

Yet another tradeoff problem arises when a skill involves both speed and accuracy. When a task like the fencing thrust requires precision movements to a defined target (accuracy) within time limitations (speed), should early learning experiences emphasize speed, accuracy, or both? Why?

_____

_____

_____

_____

_____

At one time educators felt strongly that skills should be rehearsed slowly. They contended that it was better to learn the correct movements first and that speed could be applied effectively later. However, present-day evidence suggests that a skill should be learned in the manner in which it will be performed. If a task demands both speed and accuracy, practice should reflect these demands. Speed-oriented tasks are practiced quickly, accuracy-oriented tasks more slowly.

Remedial strategies may be appropriate if a student is having difficulty in some aspect of the task. For example, if he is unable to master the fencing thrust, he could practice more slowly. You should always remember that if students practice in a manner that deviates from correct performance standards, they will still have to learn proper final skills. Readjustments will be necessary, and the learning process may take longer. Also, don't allow students to become too dependent on situational modifications, which may appear to be learning steppingstones, but which become crutches. A crutch is an effective short-term remedy, but if used too long, it may be difficult to cast aside. This leads us to still another tradeoff issue: that of real versus simulated conditions.

A simulated condition possesses many of the characteristics of the real one. It is a means of more rapidly acquiring insight into the nature of the task to be performed under real conditions, a means of practicing when performing in the real situation is not possible, or a means of concentrating on and remedying defects in performance. The simulated car or plane can serve in any of these three capacities. Ball-throwing machines in tennis simulate an opposing player and permit the student to concentrate on his stroke. In baseball they simulate the pitcher and permit the batter to concentrate on his swing.

Many other simulators may be used in sports for any of the purposes stated above. However, they may prove detrimental in the long run if depended on too long. In the decision-making strategy, consider the need for participation in real activity, and substitute simulators if they are available and can make a contribution to the student that he cannot otherwise realize.

A final tradeoff issue is that of individual versus group practice. For example, students could practice swimming strokes by themselves, or pair off in order to help one another, one playing the role of teacher, the other the role of student. If the students possess sufficient knowledge, maturity, and ability to communicate, pairing off might lead to increased motivation, reinforcement, and feedback. In order to make appropriate decisions, you must evaluate social learning situations according to the class objectives, learning theory, and research.

Information Feedback Whereas reinforcement can be any stimulus that encourages the student to repeat a response and in many cases informs him how he is doing, information feedback refers more solely to the latter process. An event can be reinforcing but not particularly informing. And an event can be informing without being reinforcing. Feedback has been used interchangeably with a number of other terms, the most common of which is knowledge of results (KR). Researchers and theorists have argued over similarities and differences among the terms information feedback, KR, reward, and reinforcement, but rather than dwelling on a semantic entanglement, we will interpret the term information feedback as broadly as possible to include related concepts.

Whenever a student performs, assuming he is aware of the model of ideal behavior, a discrepancy exists between the actual and the ideal as long as the ideal is not achieved. Thus information feedback occurs, or can occur, when this discrepancy exists. It is response-produced information. The student can receive feedback

1. from internal or external sources
2. concurrently or terminally
3. immediately after performance or delayed

Internal sources refer to the student's own means of determining the adequacy of his performance. He can see where the golf ball goes, and he feels the execution of a gymnastic movement. The visual and kinesthetic senses, individually or in combination, usually operate in the performance of athletic skills to provide the student with internal sources of feedback. Sometimes this information is sufficient so that the student can make necessary adjustments, sometimes not. It is more than likely that the beginner will not be able to use these forms of feedback to maximum advantage. In such cases external types of feedback are necessary and appropriate.

Another name for external feedback is supplementary, or augmented, feedback. Probably the most common example is the instructor's comments to the student as to the nature of his performance. Videotape would be an excellent example of supplementary feedback. In any situation supplementary feedback can be used in many ways to help the student analyze his behavior.

Depending on the nature of the activity in which the student is engaged, feedback may be obtained concurrently or terminally. If the activity is not too fast-paced, the student will be able to determine the effectiveness of the individual behaviors of which the entire activity is composed. After it is completed, he can reflect on what has transpired. Similarly, deliberate attempts can be made to induce concurrent (ongoing) feedback through prompts and cues and certainly to provide terminal feedback when the task is completed. The desirability of either or both will depend on the nature of the student's behavior and skill level.

Finally, feedback can be administered immediately or delayed. For example, in a volleyball class the teacher could appraise each student's serve right away or wait until the end of the class to do so. Some form of feedback is better than none, and generally speaking, it should be immediate rather than delayed.

Cue Generalization and Discrimination   Learning situations involve a process whereby the student at first attempts to generalize from previous experiences, then acquires task- or behavior-specific discrimination abilities, and finally learns to generalize to other, related situations. Generalizations lead to discriminations and then to further generalizations. You will recall that in the GAMES approach the E stands for Experienced for transfer effects and that there is continual transfer of learning from previously acquired behaviors to those to be acquired. Due to dissimilar previous experiences some students are more likely than others to succeed.

An ability to generalize from past to present experiences is a decided advantage for the student in the early stages of learning. Whether it takes the form of rules, tactics, stimulus agents, or movement patterns, generalizing from the previous to the present can occur if the student (a) possesses the prerequisite knowledges and skills, and (b) can see relationships and make associations. Laboratory research continually indicates that students not only learn to respond to the specific stimulus but to related ones as well, and that closely related stimuli elicit stronger generalizations. Previously learned tennis skills will probably generalize reasonably well to paddle ball or racket ball skills; and to a lesser degree, to squash, badminton, and handball.

Generalization enables the student to achieve well at first. However, he can demonstrate skills and certain forms of cognitive material proficiently only after he has learned to discriminate among cues effectively. Task-specific practice helps him develop discrimination. In order to excel at his specific endeavor, the student learns how to distinguish among closely allied cues, to filter out irrelevant and meaningless ones, and to respond with precision and timing. Thus, effective cue discrimination and response adjustments are associated with highly skilled performance.

Often learning situations are not ends in themselves but rather means for further learnings. In other words, classes and other educational experiences can be managed to promote discriminating and generalizing abilities which will meet present demands as well as facilitate achievement in related future endeavors. Class objectives should indicate the extent to which these abilities should be developed in terms of psychomotor, affective, social, or cognitive behaviors.

Event Repetition Students must have sufficient opportunity for practice and experience either in or outside of class if they are to improve their performances and behavioral potential. Extensive practice is necessary in learning highly specific skills, yet an attitude or concept can be acquired or modified as a result of one situation. Thus the nature of a particular objective and its competency level will dictate the extent of practice or experience necessary for its realization.

A perfect tennis serve, the ability to execute numerous sit-ups, and a complete understanding of golf terminology and rules are attained through repetition. On the other hand, a particular attitude may be modified during one or a series of well-conceived situations. Attitudes toward sportsmanship, interest in participation, and social and racial relations are pertinent to the physical education class. Tactics and strategies or principles of movement can be acquired in a similar way. Excessive repetition is appropriate when a well-defined response is to be associated with a specific stimulus cue. Task-specific practice leads to proficiency in that task. More general behaviors (of any kind) associated with more general situational cues need not be repetitiously learned. In fact, experiences in a wide range of circumstances will tend to lead to more flexibility in making generalizations. Attitudes, social behaviors, strategies, concepts, and general movement patterns can probably be effectively taught in a variety of situations. Above all, always consider the quality of the learning experience in relation to the time spent.

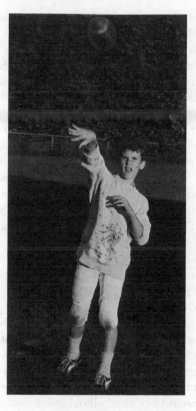

"Students must have sufficient opportunity for practice and experience either in or outside of class if they are to improve their performances and behavioral potential."

The GAMES plan described the preparatory state of the student prior to an activity, and the PRACTICE plan, the learning conditions he encounters once he has embarked on the task. To review, PRACTICE includes Practice, Reinforcement, Administration, Cue generalization and discrimination, Tradeoffs, Information feedback, Contiguity, and Event repetition. Practice can lead to perfected skills and more desirable behaviors if PRACTICE conditions are considered.

The Problem of Retention

So far, we have reviewed learning processes and conditions appropriate for the acquisition and modification of behaviors. Little thought has been given to the ways of promoting the retention of what has been learned. Yet we are often required to demonstrate what we know after a period of time has elapsed since our initial practice or experience. When we are tested on the material covered in a course such as this one, how well will we remember the essential points? How much will we forget? Will the motor skills displayed on one test be replicated on the next, or will they be forgotten to some extent?

Retention refers to the amount of tested material remembered, and forgetting, obviously, to the material lost over a period of time. Earlier in this chapter we stated that learning results in a relatively permanent change in behavior. By implication, then, the better something is learned, the more effectively it is retained. When developing the content for the classes you will teach, you should realize that ultimately the relative amount of high-quality time spent on each phase and on certain aspects of the content, as well as the instructional strategies employed, will greatly influence the probability of retention. The length of time between the last period of practice and the test or new attempts will also affect retention. So will whatever occurs during this elapsed time. Let's examine some plausible explanations of the memory process and formulate practical ways of encouraging retention.

Theories of forgetting have primarily tended to fall into two categories: *disuse* and *competing responses*. Advocates of the disuse concept hold that forgetting occurs when activities are not practiced and repeated; performance declines with the passage of time. The theory of competing responses suggests that the nature and quantity of events that occur over time will affect the retention of once-learned material. You might think of this theory as one that embraces the concept of negative transfer. Recently experienced activities can interfere with the retention of previously learned ones, a condition known as retroactive inhibition. For instance, if you had learned to play golf last year, and if during the present year baseball occupied your time and you allotted no practice to golf, it is conceivable that certain golf skills would be weakened through the process of interference.

Other explanations of forgetting are quite dissimilar. There are those who believe that once something is learned, it is never truly forgotten; Freud's concept of repression suggests that a person does not really forget unpleasant experiences but rather resists recalling them. Recent work in biochemistry has demonstrated the importance of RNA (ribonucleic acid) to memory. Evidently, learning produces chemical changes in the brain which help to explain why and how things are remembered. Reorganization theory states that memory is a function of the ability to generalize from past to present experiences. An inability to reorganize processes inhibits the performance of present activities and indicates poor retention. Finally, short- and long-term retention have been studied in great detail in recent years, and resulting data support the theory that the human organism can process just so much information at one time. The stimulus situation must be of sufficient duration, without an overload on the processing system, so that the information can pass from the short-term to the long-term retention stage. Storage and retrieval are thus a function of exposure time.

Considering all these explanations of forgetting, what can we do to facilitate retention? What conditions should be identified and manipulated?

1. *Overlearning.* Extensive practice or experience (stimulus-response associations) tends to fortify bonds and encourage retention. Overlearning discourages competing responses from negatively influencing performance on the task at hand.
2. *Meaningfulness.* Meaningful material is retained better. The student should realize the relevance of the tasks to be learned. If he enjoys and values them highly, he will probably remember them longer than less meaningful material.
3. *Serial order.* The order in which matter is presented to the student, especially verbal materials, will affect retention. Beginning and terminal materials are usually remembered best; middle ones least well. Sequential arrangements should be considered in this light. The advantage of part practice over whole practice should be considered here, for deliberate part practice on different aspects of the task might minimize the serial-order effect of whole practice. This is especially true with extensive lists or other verbal materials.
4. *Intervening activities.* The nature and extent of the materials to be experienced following original learning and later recall should be controlled. The original task and newly introduced tasks might be practiced alternately. Or the type of tasks subsequently learned might be carefully selected to minimize retroactive inhibition. Practice time might also be controlled.

Summary

At this stage in your reading, you should be able to define an instructional strategy as well as identify and describe the four event variables underlying its development: type, scope, sequence, and presentation. Objectives and student entry behaviors influence instructional content. You should analyze the learning event in terms of internal and external conditions, i.e., the student's readiness and willingness to learn, and the instructional conditions imposed by the teacher. Once you have decided on the type and sequence of events, determine the scope and manner of presentation of each. Also, it is important to note that sequencing can be applied to the ordering of learning events as well as to the relationship of a hierarchy of tasks geared to achieve a particular objective.

You should also be aware that learning is typically inferred from behavior or performance, and connotes a relatively permanent change due to practice or experience. Behavior is affected not only by learning. Temporary performance states, maturation, and reflexes also influence behavioral expressions.

Learning considerations may fall into two categories: those related to the preparation of the student for learning, and those involved in the actual learning experience or practice. The GAMES plan (Goal orientation, Attention, Motivation, Experienced for transfer, and Situational familiarity) offered guidelines for the first category, and the PRACTICE plan (Practice, Reinforcement, Administration, Contiguity, Tradeoffs, Information feedback, Cue discrimination and generalization, and Event repetition) suggested guidelines for the second. All the material presented in this chapter contributes to the appropriate development of instructional strategies, which you should now be ready to formulate.

Resources

Books

Education

Baird, Hugh, Belt, W. D., Holder, Lyal, and Webb, Clark. *A Behavioral Approach to Teaching*. Dubuque, Iowa: Brown, 1972.   Each of fifteen instructional topics are analyzed according to behavioral objectives, ideas to be learned, preassessment, learning activities, and evaluation. All chapters will be of interest to physical educators, although sections 3.3 and 6.2 deal specifically with psychomotor activities.

Bandura, Albert. *Principles of Behavior Modification*. New York: Holt, Rinehart and Winston, 1969.   Extensive summary of the way various modeling cues can influence many forms of behavior.

Briggs, Leslie J. *Sequencing of Instruction in Relation to Hierarchies of Competence*. Pittsburgh: American Institutes for Research, 1968.   Technical but informative review of research literature and interpretations for instructional decisions.

Gagné, Robert M. *The Conditions of Learning*. New York: Holt, Rinehart and Winston, 1970.   Presents types of learning and learning events. May be a classic in its time.

Gerlach, Vernon S., and Ely, Donald P. *Teaching and Media: A Systematic Approach*. Englewood Cliffs, N.J.: Prentice-Hall, 1971.   A number of chapters contain material useful in understanding and applying the contents of our Chapter 8.

Haney, John B., and Ullmer, Eldon J. *Educational Media and the Teacher*. Dubuque, Iowa: Brown, 1970.   Practical approaches to shaping learning.

Merrill, M. David. "Psychomotor and Memorization Behavior." In M. David Merrill (ed.), *Instructional Design: Readings*. Englewood Cliffs, N.J.: Prentice-Hall, 1971.   Easy-to-follow explanation of relation between psychomotor behaviors and learning conditions and process.

Merrill, M. David, and Goodman, R. Irwin. *Instructional Strategies and Media: A Place to Begin*. Monmouth, Ore.: Teaching Research Division, Oregon State System of Higher Education, 1971.   Although in rough form, contains good ideas and techniques for improving psychomotor, cognitive, and affective behaviors.

Twelker, Paul A. "Designing Instructional Systems." In Jack Crawford (ed.), *CORD National Research Training Manual*. Monmouth, Ore.: Teaching Research Division, Oregon State System of Higher Education, 1969. This chapter, as well as others, deals with aspects of a systems approach to instruction.

Physical Education

AAHPER. *Organizational Patterns for Instruction in Physical Education*. Washington, D.C.: AAHPER, 1971. Practical and innovative ideas and organizational patterns for organizing time allotment, personnel, and student grouping structure in physical education.

Mosston, Muska. *Teaching Physical Education*. Columbus, Ohio: Merrill, 1966. Alternative strategies for reaching goals, and recommended ones with good examples.

Singer, Robert N. *Coaching, Athletics, and Psychology*. New York: McGraw-Hill, 1972. Analysis of learning research and theory as applied to coaching and athletics.

Singer, Robert N. *Motor Learning and Human Performance*. New York: Macmillan, 1968. Learning research and theory as related to motor skills of all kinds.

Articles

Education

Tosti, Donald T., and Ball, John R. "A Behavioral Approach to Instructional Design and Media Selection." *AV Communication Review*, 17: 5–25, 1969. Creative approach to instruction and presentation modes.

Physical Education

Dougherty, Neil J. "A Plan for the Analysis of Teacher–Pupil Interaction in Physical Education Classes." *Quest*, Monograph XV:39–50, 1971. Discusses the nature of verbal interaction between the teacher and students. Other excellent articles are also found in this issue, e.g. "Educational Change in the Teaching of Physical Education."

Gentile, A. M. "A Working Model of Skill Acquisition with Application to Teaching." *Quest*, Monograph XVII:3–23, 1972. Applies knowledge of the process of motor learning to the teaching of skills. Other articles in this issue, e.g. "Learning Models and the Acquisition of Motor Skill," are also worthwhile.

9
Using Media

In this chapter you will become familiar with various types of media for use in both instruction and evaluation. Guidelines for their selection and implementation will be suggested. The functions of media in specific areas of instruction are discussed, and an analysis is made of research findings concerning the role of media in physical education. A teacher selects a particular instructional medium because he feels it will best enable his students to attain their objectives. He wants to optimize the learning experience. In order to do so, however, he must state objectives clearly, analyze media characteristics, and sequence the appropriate intermix of people, environment, and sources.

Chapter Contents	**Media and the Instructional Process** What Are Media? The Implications of Media Choosing Media **Media in Physical Education** Classifying Media by Domain Classifying Media by Function **Specific Media** Videotape Recorders Motion Pictures, Filmstrips, and Closed-circuit Television Simulators and Trainers Printed and Static Material Tape Recorders, Records, and Radios Automation and Programmed Materials **The Research Problem** **Summary**
Student Objectives	The student who understands the material in this chapter should, with the book closed, be able to:

1. Identify a minimum of twelve media resources for possible use in instruction
2. State three factors to consider when selecting particular media
3. Compare media associated with cognitive behaviors with those associated with psychomotor behaviors
4. Explain seven instructional functions of media, and give an appropriate medium for each
5. Describe the functions of the following media in physical education and discuss related findings:
   a. videotape recorders
   b. motion pictures, filmstrips, and closed-circuit television
   c. simulators and trainers
   d. printed and static material
   e. tape recorders, records, and radios
   f. automation and programmed materials
6. Analyze the general status of research on media in physical education, giving two major reasons why it is difficult to apply research to practice

Media and the Instructional Process

In Chapter Eight we discussed learning strategies and instructional media. Instructional objectives can invariably be met to some extent with alternative strategies and presentation of content. Since we are attempting to analyze each teaching situation in terms of appropriate techniques and content, it seems necessary to explore further the possibilities of instructional media.

What Are Media?

In physical education the basic medium is the gymnasium. For a unit in basketball additional media might include court constraints, necessary equipment, and the teacher. In the usual medium for teaching physical education activities, the teacher models and guides behaviors, and the students' experiences occur under "real" conditions; that is, actual practice and games on the basketball court. But there are alternative procedures.

The medium can be structured or restructured to include various kinds of technological and other instructional aids or media, and simulated conditions—not merely for the sake of change but also in the interest of better instructional methods and outcomes. Media is a term familiar to all of us. We usually associate it with films, charts, and records, but many other forms are used in contemporary education. At one time media were traditionally referred to as audio-visual aids, with a heavy emphasis on "aids." In other words, media were thought of as supplementary materials. Today, however, it is obvious that media alone can be used to

"In physical education the basic medium is the gymnasium."

instruct students. For instance, when you use closed-circuit television and programmed learning methods (texts, machines), you are no longer the main source of information for your students.

Media is the term we will use in this chapter for materials and equipment you may use to replace partially or supplement your usual instructional procedures. You can also use them to evaluate students quickly and accurately. Media are popularly bisected into hardware and software configurations. At any rate, an examination of our systems model (Figure 9-1) reveals that we are nearing the end of our instructional considerations as we analyze the role and purpose of media in the physical education program.

The Implications of Media

Advances in technology and a growing concern for individual learning rates and styles have created a media boom in education. Every year a greater number and variety of traditional instructional materials are produced in the form of textbooks, films, charts, and the like. Coupled with newly developed media, they make special and difficult demands on the teacher, who must decide what to use and when and how to use it. Videotape, closed-circuit television, teaching machines, programmed texts, and computers have introduced unbelievably varied instructional possibilities. The teacher is becoming increasingly aware of the diversity of

FIG. 9-1 *Systems approach model*

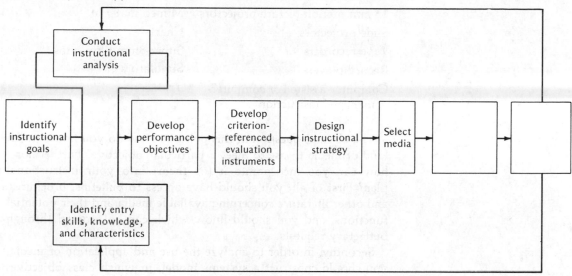

media as well as of his ignorance of its effective use in the instructional process.

Most media resources are designed to meet cognitive and behavioral objectives. They can also enhance the cognitive aspects of motor performance, such as goal direction, cueing, and feedback; and can be used to transmit rules, equipment and facility specifications, terms, tactics, and other matter. Media can assume certain of the teacher's functions, thus relieving him to concentrate on individual problems. In the form of simulators or trainers, media can simplify skill practice. When real conditions are unavailable, the student can practice with such equipment, which will enable him to isolate and emphasize aspects of his performance and thus maintain acquired skill. Of all the potentialities of media, one of the most important is the role they can play in individualized instruction.

Choosing Media   Although it does not contain every conceivable type of media, the following list is long enough to include a variety of materials and equipment, any one or combination of which can be used in most physical education classes. Not listed are unique materials and equipment that have been developed for specific activities; examples of these will be presented later.

Videotape recorders (VTR)   Textbooks
Closed-circuit television sets   Programmed texts

Overhead projectors	Self-instructional kits
16 mm motion-picture projectors	Mimeo material
Slide projectors	Charts
Tape recorders	Super-8mm film loops
Record players	Simulators
Computer-assisted or computer-managed instruction	Trainers

The obvious question at this point is: How do you know which kind of media to use to meet a particular objective? Furthermore, how can you incorporate it (or them) into your instructional plan? First of all, you should have access to bulletins, brochures, and other literature concerning available media and their potential functions, and you should know which can be ordered through budgetary requests.

Secondly, in order to analyze the use and application of media, you should construct a systems model, in which class objectives would be specified, behaviors classified, learning events (GAMES and PRACTICE plans) determined, alternate strategies and mediums compared, specific media options (cost-benefit tradeoffs, availability, technical quality, etc.) analyzed, evaluation needs appraised, and decisions made.

Finally, you should be familiar with pertinent research and theory in order to defend your choice of a particular medium. As you will soon see, research with media and psychomotor skills has been sparse, and many results are inconclusive. Often theory and common sense might indicate the use of one medium in preference to another. Thus in selecting media, *media information and availability*, *systems analysis*, and *research and theory* will be your general guidelines. More specifically, media are best at providing external stimuli which can be used in *setting goals*, *directing attention*, *presenting content*, *cue prompting for responses*, *producing reinforcement*, *providing information feedback*, and *evaluating students*.

 The teacher of physical education plays a more prominent role than the teacher of cognitively oriented classes. The latter can use computer-assisted or computer-managed instruction, programmed texts, and various audio and visual equipment. A class in mathematics can be and has been taught almost entirely with media, which are obviously much more than instructional aids in this case. A physical education class, because of its unique content and objectives, cannot as easily be turned over to media, although they can assume some of the functions of content presentation as well as providing supplementary benefits (goal setting, cue prompting, etc.) to traditional instruction. Generally speaking, films, closed-circuit television, text material, and the like can be used for group instruction, and videotape and tape recorders for individual instruction.

Focus

Many proponents of multi-media (media in combination) may be found in education today. Leslie Briggs, for one, has investigated many of the possibilities of media and has encouraged a number of teachers to alter their conventional approaches. Together with Peggie Campeau, Robert Gagné, and Mark May, he has made a significant contribution to the literature entitled Instructional Media. *The following excerpt concerns the value of multi-media instruction.*

First, it may be said that while the spoken word and the printed word are, indeed, very important forms of instructional stimuli, they are not the most effective or economical teaching stimuli for many kinds of objectives. Graphic illustrations are a feature of most textbooks, and motions and space relations are better depicted (especially for young children) by three-dimensional models or motion pictures. Sounds other than speech need to be heard for many purposes—in music, mechanics, equipment operation. Thus, other forms of audio and visual stimuli are needed to supplement written and spoken words.

Second, in addition to selecting the sensory mode to be stimulated, verbally or otherwise, it is important to note that various educational objectives represent *different kinds of learning, for each of which the external conditions of learning may be quite different.*

Third, to supply the instructional events needed for each type of learning and to display instructional stimuli required in the sensory mode and quality needed, various media (audiovisual) devices are of differential effectiveness, depending upon the exact learning requirement imposed by the educational objective.

Fourth, there is sufficient knowledge (though yet very incomplete knowledge) as to the conditions required for the various kinds of learning, that this information, considered along with analysis of the instructional events and stimulus displays for each objective, can result in selection of a relevant instructional medium or media. These selections, translated into programing specifications, can result in the means for achieving improved effectiveness in learning.

It is important to note that as crucial as is the step of identifying the *type* of learning required by the objective, in order to specify the external condition of learning, this step does not of itself lead to immediate choice of the medium. Many other factors, discussed later, are taken into consideration. For example, a necessary condition may be to provide contiguity between two stimuli; but identifying the sensory mode, and type, duration, and intensity of the two stimuli is also needed. It may then be apparent that any of several media may be able to present these stimuli in the particular manner needed to satisfy the instructional event required. The variety of stimuli needed, primarily in auditory and visual modes (both verbal and nonverbal), suggest the value of multimedia instruction. However, for some objectives, monitoring and correcting responses are also important.

In summary, multiple media are needed primarily to provide the range of displays needed, and to provide feedback and evaluation of responses. In total, the media are the means for providing stimuli, whether the stimuli are used to motivate, to direct attention, to set a goal, to give a prompt, to evaluate, to guide thinking, to evoke a response, or to test for transfer.

To summarize the above points, the central rationale for the proposed solution to the problem of how to choose instructional media rests on the assumption that various educational objectives require different kinds of learning. These kinds of learning, in turn, are established by different sets of conditions. The sets of conditions of learning in their turn, are made possible by the manipulation of instructional events. The way of providing or producing these instructional events is to apply appropriate stimuli. The sensory mode to be stimulated, and the detailed characteristics of the needed stimuli, together are considered in order to select the mechanisms or media to be employed to present the stimuli. Media are thus vehicles for stimulus presentation. (1967, pp. 10–11)

Media in Physical Education

Much material is currently being assembled on the role and probable advantages of media in instructional settings, *but* in usual class situations. Since much of education is becoming highly mediated, we must assume that physical education will also as media are developed and are shown to benefit instruction and evaluation.

It is wise to keep in mind that media themselves do not affect the student or his learning. Whether he learns depends on the appropriateness of the media in meeting objectives, how they are presented, and his ability to respond favorably. Thus media are only effective if properly used. Doing so is not easy, for at present there is no organized body of pertinent literature for physical education. You might follow the guidelines of Donald Tosti and John Ball (1969, pp. 7–8) for selecting media in any instructional setting:

1. Determine the nature of the problem and establish general goals to solve the problem. [In physical education, the problem may be to teach tennis to a class of high school sophomore girls so that they achieve a minimum level of skills, the level of which has been pre-established.]

2. Determine the specific behaviors to be established and the entering behaviors of the students.

3. Deduce the presentation factors which produce the desired behavioral effect employing established evidence in learning; then analyze or synthesize the generalized response sets which may be employed by the student in his response to the presentation.

4. Select media which fit the presentation requirements. Media selection must be done in terms of eliminating media which limit or otherwise adversely affect the presentation design rather than specifying advantageous media. Then assemble an operational instructional systems package (media-mix).

5. Determine strategy for introduction of the operational system into the instructional environment.

These guidelines emphasize the importance of analyzing the situation carefully and then selecting media discriminately. Media for potential use in physical education can be categorized by behavioral association or function. Possible external stimulus functions have already been alluded to. Types of media would include (a) the usual equipment, material, and facilities used in the real situation, and (b) supplementary aids used in practice—visual, auditory, and kinesthetic materials, and simulation and training equipment. Both categories will be discussed here, with some overlap.

Classifying Media by Domain

The most obvious substitutes for real situations are realistic practice conditions, which contain all the elements and aspects of the final evaluation. Students practice basketball skills with official equipment and on standard courts, in a traditional teacher-student interaction approach. The medium is the gymnasium, the teacher orients and directs student behaviors, and practice proceeds with no additions to or modifications of the environment. But if media are to be added, it is convenient to categorize them according to the type of behaviors (e.g., cognitive or psychomotor) they affect.

Cognitive Behaviors Assuming the word media is broadly interpreted, materials, equipment, people, and environments in any learning situation can be intermixed in many ways. The media discussed under the heading "Cognitive Behaviors" include the usual reference material and equipment students use while they are relatively inactive and are attempting to process information through the auditory and visual senses. These media should be analyzed in terms of the possible contribution they might make in achieving cognitive objectives.

There is a great variety of visual media, which may be classified as projected or nonprojected. Closed-circuit television and motion pictures can be used to present class content and direct students' goals. Slide projectors, overhead projectors, and filmstrip projectors can serve similar functions. Since they show material in static form, they are more limited in some ways but have other uses. The videotape recorder, one of the more widely heralded media for physical educators, is an excellent source of immediate feedback. In the nonprojected (printed item and static visual) category, both conventional and programmed texts, charts, photographs, illustrations, and mimeo material can fulfill content presentation requirements and help establish goals and direct students' attention. Programmed text material (examples of which were given in Chapter Two) also provides immediate reinforcement, information feedback, and self-evaluation.

Besides film soundtracks and the teacher's voice, auditory media include records and record players, radios, and tape recorders. Auditory media can provide almost any kind of external stimulus in addition to presenting content. We are primarily considering those media which contribute to knowledge, understanding, and other cognitive components of behavior, but can also directly benefit the learning of psychomotor skills in such ways as providing prompts for appropriate responses.

Psychomotor Behaviors   The material described here can be used while the student is performing. They are geared to improve skills or, at the very least, maintain them.

Visual cues and aids in psychomotor activities vary, but any modification of normal procedures must be viewed with caution. Remember, your ultimate objective is to have your students perform well under real conditions. Visual cues (e.g., a marker placed before the archery target) can simplify the learning process. However, if they will not be present in the actual activity situation, these cues should not be used too much, for they will then become a crutch. The student will become too dependent on them and will be unable to function effectively without them. These cues should gradually be faded out after they have served their purpose.

Can you think of an activity in which visual cues might be added to the learning environment in order to promote the acquisition of skill?

_____

_____

_____

Auditory cues in motor performance are most commonly associated with phonograph records of rhythmic sounds or music which help students coordinate their movements. Tape recorders can also be used in physical education, however. The teacher might simply record his comments as each student performed, and the student could play the tape at his own convenience. Kinesthetic aids usually promote learning by simplifying a task and encouraging the student to practice a portion of it. They allow him to "get the feel" of the movement. The kickboard frees the swimmer from worrying about the arm stroke when learning the crawl so that he can concentrate on rhythmic kicking. Kinesthetic aids can also be used for transfer purposes. In ball sports, projectiles heavier or lighter than regulation weight may promote better performance by sharpening the kinesthetic sense. Weighted bats, balls, spats, boots, and such for practice in throwing, jumping, hitting, and running are sometimes thought to result in increased effectiveness immediately after use. Although the effectiveness of these aids has not been proved, they are alternatives to traditional media.

Trainers and simulators have long been used in teaching psychomotor skills. (It is often hard to distinguish between the two media, but semantics need not stand in our way.) Trainers usually show the student what is expected of him or allow him to practice skills by himself. Simulators, as the name implies, approximate real conditions. Trainers are used primarily by beginning students, and simulators by the more advanced. But in almost any sport specially designed equipment can help students perfect or maintain their skills.

These media are valuable because they usually simplify and minimize stimuli, thus enabling practice of selected skills (examples are ball-throwing machines in tennis and baseball). The teacher must make tradeoff decisions as to real versus artificial practice. Although research and theory generally indicate the value of practice in actual activity situations, there may be times when trainers and simulators are especially valuable. The absence of real practice conditions (snow on the golf course, no snow on the ski slope) can be somewhat overcome with computerized golf equipment, which simulates playing an actual round of golf, or a simulated ski slope. Students who have difficulty acquiring skill in certain aspects of an activity might develop further if they practiced with media of this kind.

Classifying Media by Function

Media can be analyzed according to function as well as by behavioral association. We will draw on a number of the learning events and strategies discussed in Chapter Eight and attempt to relate the potential contributions of various kinds of media to them (although we cannot explore all possibilities or treat functions in depth here).

Remember that a number of media can contribute to, or supplement, any learning event. Individual media can function adequately, but the multi-media approach will help insure adequate stimulation for more students. Since media in the broadest sense of the word can refer to any environmental component that communicates with the student, it is apparent that the most conventional means of instructional communication is the teacher's voice. Although oral communication effectively transmits many instructional events, it has certain limitations. For instance, if the class objective is to hit a golf ball with reasonable form (according to specified criteria), words can only go so far in helping students visualize goal expectations. Live demonstrations, films (especially with a sound track), and pictures are more capable of stimulating the student and communicating the intended goal.

Focus

Since various media can serve the same function, how do you know which medium to select and use in the instructional process? Before we can answer this question satisfactorily, we should note that research has not overwhelmingly supported the use of particular media over others in learning situations. It is important to know and be able to compare media functions. But there are other concerns. Robert Gagné presents three general conclusions from research findings in The Conditions of Learning. *You should comprehend them before you attempt to select from alternative instructional media.*

How is one to decide which media to use for which instructional purposes? This question has been the subject of much research over a period of many years. Some of the conclusions that can be drawn from this research may be valuable as an introduction to the problem.

First, no single medium is likely to have properties that make it best for all purposes. When effectiveness of one medium is compared with another for instruction in any given subject, it is rare for significant differences to be found. Lectures have been compared with reading, lectures with motion pictures, pictures with text, and many other kinds of comparisons have been made without revealing clear superiority for any given medium. At any given time, a medium may enjoy unusual popularity, as has been the case, for example, with television, teaching machines, and computerized instruction, at one period or another. Sometimes one medium is found by research to have an advantage for one subject matter only to be shown to have none for some other subject matter. Over a period of years, researchers have learned to be skeptical of single instances of reported statistical superiority of one medium versus another.

Most instructional functions can be performed by most media. The oral presentation of a teacher can be used to gain and control attention, but so also can the use of paragraph headings in a textbook, or an animated sequence in an instructional motion picture. The learner can be informed of the expected outcomes of instruction by a printed text, by an oral communication, or in some instances by a picture or diagram. Recall of prerequisite learned capabilities can be done by oral communication, by means of a sentence or picture in a text, or by a movie or television pictorial sequence. Similar remarks could be made about every one of the functions of instruction described in this chapter. It is possible, of course, that additional research of an analytic nature may yet reveal some important special properties of single media that make them peculiarly adapted to one instructional function or another. Up to now, however, the most reasonable generalization is that all media are capable of performing these functions.

In general, media have not been found to be differentially effective for different people. It is an old idea that some people may be "visual-minded" and therefore learn more readily from visual presentations, while others may be "auditory-minded," and therefore learn better from auditory presentations. While a number of studies have been conducted with the aim of matching media to human ability differences, it is difficult to find any investigations from which one can draw unequivocal conclusions. If this idea has validity, it has not yet been demonstrated. A possible exception is this: several studies have shown that pictorial presentations may be more effective than printed texts for those who have reading difficulties or small vocabularies. This is hardly a surprising result, and it seems wise to refrain from overgeneralizing its significance.

In view of these findings, which serve the purpose of "clearing the air" in discussing media or in making the most effective use of available media? The suggestion to be made is that decisions about media should probably be made in sequential stages, choosing the best alternatives at each stage. (1970, pp. 363–365)

Let's examine an instructional function such as goal setting, and associate various media with compatible properties. We are not evaluating the capacity of each medium to achieve particular functions, but merely citing those that can function acceptably in specified ways. The actual selection of a particular medium will depend on a number of factors, such as (1) clearly and properly stated objectives, (2) an understanding of the characteristics of the medium, and (3) the ability to integrate it into an effective instructional sequence.

But one major consideration is the instructional objectives established by the teacher. What will the students be expected to do? How? In what context? If students are expected to perform psychomotor skills, they should observe dynamic examples of the desired behavior. If they are supposed merely to explain the components of an act, perhaps oral or written communications would suffice. The following list contains examples of learning events and the possible role of media.

Functions	Media
1. Setting Goals (the student knows what is expected of him)	Oral communication Films Slides Printed matter Charts, drawings Photographs Demonstration Closed-circuit television
2. Directing Attention (the student is alert and receptive)	Almost any media
3. Motivation (the student is optimally stimulated to perform)	Almost any media
4. Presenting Content (the student is provided with appropriate information)	Almost any media
5. Prompting Correct Responses (the student is furnished with external prompts)	Oral communication Training devices Records Printed material
6. Information Feedback and	Workbooks Programmed texts Oral communication Videotape recorders Tape recorders
7. Reinforcement (the student is informed of the correctness of his response)	Printed matter Teaching machines
8. Inducing Transfer (the student is shown the relationship of and can benefit from prior experiences and learning to the present task)	Oral communication (other media, e.g., printed matter, to a lesser degree)

9. Tradeoffs   Oral communication
   (speed-accuracy, part-   Simulators and trainers
   whole, etc.)

10. Evaluation   Videotape recorders
    (the student determines the   Programmed texts
    correctness of his   Simulators
    performance)

Media and learning events should be considered within the framework of behavioral objectives. In Chapter Five we categorized behaviors into four domains—cognitive, affective, psychomotor, and social. Thus a specified, described, and stated behavioral outcome enables you to specify instructional events (the GAMES plan and the PRACTICE plan) and the appropriate medium for each. We just reviewed the possible functions of media, but it should be emphasized that we did not delineate each function according to a stated behavior or instructional objective. Expressing the desired function of a medium precisely enables you to choose it with more assurance.

One additional thought on the use of media: It is usually wise to test students in the same medium in which they have been trained; making testing situations dissimilar to learning situations is unfair. Furthermore, the validity of the data will be questionable. Exceptions to this general rule might be necessary when special media are required for students demonstrating unique problems or when media play a special role in enriching experiences.

Specific Media

Unique group or individual characteristics must always be considered in the context of research with standard samples of subjects. For instance, it should be fairly obvious that younger students and those who read poorly will not do as well as more mature and intelligent students with printed material. Oral communication, illustrations, and/or demonstration will mean more to the former. Independent and bright children would probably fare better than others with self-paced and programmed materials.

Leslie Briggs (1970) has stated these and other considerations for instructional development. He wisely summarizes by saying:

> The above conclusions reported for particular groups of students for particular subject matter prepared in particular media are offered for consideration, *not* for blind application. Learner characteristics should be considered in conjunction with task characteristics.
>
> The course designer may be aware only of gross characteristics of the students for whom the materials are being prepared, such as age, I.Q., and prior competencies. The classroom teacher using the materials for a particular group of students will, of course, have more knowledge of

each child after several weeks of his class. Often the teacher will either use the prepared materials somewhat differently for different pupils, or he may seek alternate materials for particular portions of the work for some students. The teacher can also modify the *effects* of prepared materials by grouping materials. (pp. 95–96)

With these thoughts in mind, let us turn to major media possibilities in physical education. By major media we mean those which appear to have the greatest potentiality for physical educators, apart from conventional teaching methods such as lectures.

Videotape Recorders   Of all the media developed in recent years, the videotape recorder (VTR) seems to be the most exciting and meaningful for students learning psychomotor skills. The student performs in front of a television camera and can then observe his execution immediately afterwards or at any other time, since a permanent record has been made of it. Often we do not realize the way our performance really looks. Imperfect movements are probably communicated more effectively to students with the VTR than by means of verbal communication.

Researchers have fairly well established that feedback is effective when immediate and specific. The feedback modalities available to the student differ from skill to skill. In sports like basketball, archery, baseball, and golf, the student can feel and see how he has done. The sense of feel, or kinesthetic feedback, is what the gymnast or diver relies on. Even in those sports in which the performer is provided with visual feedback about the results of his execution, he may need more information about it. For this reason VTR is being adopted increasingly by skiing, golf, and tennis instructors, varsity and professional athletic teams, and physical education teachers. Field reports and general appraisals are highly enthusiastic.

Some degree of natural feedback, in the form of vision and feel, is present in most activities. Videotape is a form of artificial, or supplementary, feedback. It informs the student of how his performance looked, rather than the consequences of it, although consequences are often directly related to manner of execution. In order to interpret the research meaningfully, we must examine studies dealing with feedback and knowledge of results, since these terms have been used somewhat interchangeably in the past. Furthermore, we might analyze the process of acquiring knowledge of results, or feedback. Learning feedback is given after a sequence response is made; performance feedback during the actual sequence of responses. Videotape is obviously a form of learning feedback.

Research has traditionally compared groups of students, one learning an activity with the use of VTR and the other without

"Researchers have fairly well established that feedback is effective when immediate and specific."

it. At first glance the results are disappointing; usually, no differences in performance are noted.

But remember that in Chapter Five we classified skills as being self-paced or externally paced. Self-paced tasks emphasize form, style, and precise movement, and rely heavily on biomechanical principles. VTR probably contributes more to them than to externally paced tasks, in which a variety of responses are acceptable as long as the desired result occurs. If used correctly, VTR can also help students learn externally paced tasks, but probably to a lesser degree.

In order to make the most of media, you might show films or cartridge tapes of ideal performances along with the VTR tape of the individual student's performance. The student will then have a reference model. Instructional technologists recommend using a medium that enables the student to compare his actual performance with the desired performance. This process also allows students to learn independently of the teacher and at their speed. It may also prove to be a good means of incentive. The VTR is a well-designed evaluative technique for students.

As a final consideration, the VTR will be effective if the students are mature enough to know how to analyze their movements and if they have been taught how to do so. It takes a critical eye, knowledge, a willingness to analyze, and an ability to transform visual images into effective response patterns.

Focus

We present here a representative study, "Video Feedback in Learning Beginning Trampoline" by Pamela E. James, of the use of videotape in a physical education setting.

In some skills which require the linking of precise or finely controlled items, visual feedback of results is not available to the learner. Skills in sport involve activities of this kind, where such feedback is usually provided by verbal communication through speech. However, visual feedback can be supplied if performance is recorded on film or videotape. Using videotape, the learner can see an instant playback of his actions, thus eliminating the time necessary to develop and process film.

To investigate the effect of visual and verbal feedback in learning beginning trampoline, 18 11- to 12-yr.-old boys were assigned to two groups, matched for performance on this skill by two independent assessors and general physical ability as rated on a 5-point scale. At separate times, each group received equal instruction, i.e., 1 hr. twice weekly for 11 sessions. Subjects learned four basic drops and a seven-bounce routine. Group V (visual) was shown visual feedback of performance (using videotape) on a 23-in. monitor. One camera in a fixed position was linked to a videotape recorder. The latter was placed by the trampoline and operated by the instructor. Group NV (non-visual) received verbal feedback only. To determine the effect of verbal ability in the interpretation of feedback the Mill-Hill Vocabulary Test, Form 1, Junior was given.

Although Group V (Mean = 20.5) scored higher than Group NV (Mean = 19.5) on the four basic drops and on the seven-bounce routine (Group V, Mean = 6.12; Group NV, Mean = 5.1), this was not significant at the .05 level. A correlation between scores on performance and verbal ability in Group NV indicated that subjects with high verbal ability benefitted from verbal feedback. Boys of both high and low verbal ability benefitted from visual feedback.

Visual feedback has directive and incentive value, as the relationship between actual and desired performance can be seen. Critical self-assessment can be developed with the aid of interpretation from the instructor. Finally, more effective and efficient feedback may be supplied to learners with a wide range of verbal ability using the visual channel.

Motion Pictures, Filmstrips, and Closed-circuit Television

Motion pictures and filmstrips, in the form of training films, have been used in physical education classes, as well as military and industrial training programs for some time. Closed-circuit tele-

From James, Pamela E. "Video Feedback in Learning Beginning Trampoline," *Perceptual and Motor Skills*, 32:669-670, 1971. Reprinted with permission of author and publisher.

vision is primarily for classroom learning, although it has the potential for communicating portions of any class content to students.

The presentation of knowledge in any medium requires the student to be somewhat passive. Besides the media under consideration here, oral communication, actual demonstration, printed matter, tape recorders, loop films, slides, illustrations, and the like can serve the following purposes:

1. Inform students of the nature of the activities to be learned and ultimate goals
2. Present the content or knowledge to be learned
3. Demonstrate the significance or organization of the content
4. Guide learning, especially in early stages

When selecting media, remember that some present only visual stimuli, some only auditory stimuli, and some both. Even when they lie within the same sense modality, media can supplement each other's functions. Always consider such things as (1) the characteristics of the students, (2) the nature of the content or tasks to be mastered, (3) the desirability of group- or individual-oriented instruction, (4) the learning environment, and (5) the availability of various media.

The materials used with motion pictures, filmstrips, and closed-circuit television are carefully planned and edited. These media can (1) present content without "live" instruction, (2) rapidly disseminate information, (3) reach a large number of students, (4) show objects or people in motion, and (5) serve as well as, if not better than, most other forms of media if well-organized and clearly presented. However, filmed instruction that does not allow for two-way communication between teacher and students, can be somewhat costly if specially produced, and is not usually used in self-paced learning situations.

Films used in the military, the classroom, and industry have been extensively analyzed in terms of such variables as color, speed, audiences, content, and length. Assuming that the technical quality of these films was good, research findings have usually been reasonably supportive of them for any or all of the purposes previously mentioned.

The use of films in physical education classes has not usually resulted in better student performances. However, few studies have been reported on this topic, and possibilities have not been systematically examined. There are a number of experimental methodological concerns about past research on the possible value of any type of media, but these will be discussed later under the heading, "The Research Problem."

The quality of the film, its specific role and place in the instruc-

"The presentation of knowledge in any medium requires the student to be somewhat passive."

tional situation, the nature of the content or activity to be learned, and the audience for which it is intended will all influence students to some degree. Rather than relying on generalizations from physical education research, it might be wiser for you to analyze your situation according to all the variables mentioned, and then decide on the appropriate medium.

Simulators and Trainers

Simulators and trainers are primarily intended for the simplified practice of psychomotor tasks. Commercially or privately made equipment and materials have been widely accepted by instructors.

In physical education, training devices are one aspect of the entire range of student expectations. Ball-throwing machines in baseball and tennis permit the student to concentrate on the grooved stroking of a ball which is tossed at a consistent speed and at a designated area. Students will not have many such opportunities in reality. But mastery of fundamental skills facilitates achievement in varied and unpredictable complex situations, assuming the student makes a successful transition from the beginning to the more advanced stages of learning.

With simulators, an attempt can be made to represent the real situation, or at least a portion of it. Apparatus need not be the only means of providing simulation. Simulation of the decision-making processes in a game situation could involve simulated players and a playing area, such as a board or chart. In fact, "games" or "gaming" often represent real situations.

One of the outstanding features of simulators is that they permit the user to exert control and plan cue variation, which of course he cannot do in the real situation. The speed and direction of a ball can be controlled and manipulated with a ball-throwing machine, as can a moving target's rate of speed and distance from the archer. To review, we have identified *representativeness*, *simplification*, and *control* as outstanding characteristics of simulated material and equipment.

Simulators can be used in two major ways, for training or for evaluation. In a chapter entitled "Simulators" in Robert Glaser's, *Training Research and Education* (1962), Robert Gagné makes the following cogent comments concerning training and the use of simulation:

> Only in the case of motor skills are human activities encountered which appear to require highly accurate simulation from the beginning of learning. The weight of evidence suggests the great importance of exact task simulation from the very beginning of motor learning. Presumably, this is because many of the controlling stimuli in motor skills come from the individual's own muscles, and can therefore only be provided by requiring the movements which are to become habitual. (p. 234)

Simulators can play a role in assessing performance as well, assuming the reliability of the media and students' performances, resemblance to the real situation, and the like. Some indication of student progress, in the form of hard data, can thereby be obtained.

Research has shown that direct practice is preferable to simulated experiences, assuming the practice is of high quality and the skills to be learned are not overly complex. Activities that contain frightening or dangerous elements (swimming, trampolining) might be more suited to simulation in early learning stages. Also, individuals evincing difficulty in skill development may require task simplification and training on specific routines, which simulators are designed for. Finally, when the natural environment and regulation equipment are not available, simulated practice can be quite helpful in maintaining skill levels. Hitting perforated golf balls in the gymnasium in winter can substitute somewhat for being on the golf course.

Focus

The following article, "Relative Effects of Two Methods of Teaching the Forehand Drive in Tennis," by William H. Solley and Susan Borders, conveys an idea of how research on simulated practice has been conducted.

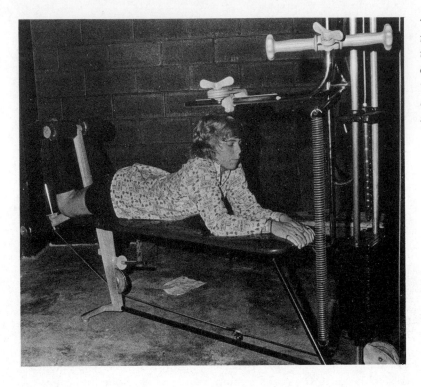

"One of the outstanding features of simulators is that they permit the user to exert control and plan cue variation, which of course he cannot do in the real situation."

This study was made to determine the effectiveness of the Ball-Boy —a machine which projects tennis balls toward the learner at regular intervals with controlled speed and direction of trajectory—in teaching a fundamental tennis stroke to a class of beginning tennis students when such effectiveness is compared with traditional teaching methods. Specifically, the study compared student progress in the forehand drive when taught by the traditional method and when taught by the traditional method supplemented by use of the Ball-Boy. Both the control group, under the traditional method, and the experimental group, adding practice with the Ball-Boy, were taught by demonstration, explanation, practice, and individual corrections. Each student in the experimental group stroked a minimum of 20 balls per class on the forehand side which where projected by the Ball-Boy.

The forehand drive, as studied, included strokes requiring directional movements toward the net and toward the sideline on the forehand side as well as strokes requiring no gross directional movements. One significant aspect of this study is whether or not enough practice trials can be given to the normal-sized tennis class to make the machine a valuable aid to teaching.

Procedures The subjects in this study were women enrolled in three beginning tennis classes in the required physical education program at the University of Florida. A combination of the matched groups and rotated groups method was employed in controlling factors other than the experimental variable which might influence achievement in the forehand stroke. . . . After orientation periods, the initial status in the forehand drive of each subject was determined. Each of the three classes under study were then divided into two groups which were matched on the basis of this initial ability.

All classes were taught all the fundamental elements of tennis throughout the study, although periodic measurements for the purposes of this study were limited to the forehand stroke. Class I received 14 periods of instruction, and the control and experimental emphases were reversed at the end of the seventh period. Each of these periods was one hour in length. Classes II and III were taught for a total of 15 periods. The initial emphasis was reversed at the end of five periods, and the new emphasis was maintained for the remaining ten periods of the study.

Classes met two days per week, either on Monday and Wednesday or Tuesday and Thursday. . . . A revision of the Broer-Miller Tennis Test was used to test ability in the forehand stroke at the beginning of the study and at all other measurement periods. . . .

Measures of forehand skill were made at the beginning of the study and at the end of each 7 hr. of instruction in Class I, and at the beginning and at the end of each 5 hr. instruction in Classes II and III.

Results In Class I, Group A, which had 7 hr. of initial instruction by the traditional method followed by the same number of hours utilizing the Ball-Boy, made considerably more progress than did Group B, which had the emphasis reversed. The difference between mean forehand skill sources of the two groups at the end of the study was significant at the 5 percent level. The same general results were noted in Classes II and III, where Group A received the traditional emphasis for 5 hr. followed by 10 hr. under the traditional method. The mean difference between the two groups at the end of the study was significant. . . .

This evidence indicates that if either of the two methods studied is to be followed by the other, it is more desirable to use the traditional method in the early class periods followed by implementation with the Ball-Boy.

Group A in all classes showed the greatest gains in forehand skill in the later stages of the study, at which time the Ball-Boy was being used. Greatest gains were also made in Group B while using the Ball-Boy, although the gains were not as large. Little or no gains were observed for students who initially learned by the Ball-Boy method and then changed to the traditional system. The complexity of chasing balls, stroking fewer balls per period, stroking balls requiring many footwork patterns, stroking balls thrown or stroked to them inaccurately, and many other variables in the traditional method of instruction apparently caused learners to strike a plateau and often to regress from the level of achievement reached under the simpler Ball-Boy techniques. All classes showed significantly greater progress while learning by the Ball-Boy method.

Summary and Conclusions   The following conclusions were reached as a result of this study:

1. Teaching machines which aid in standardizing and controlling speed and direction of practice balls are highly valuable in teaching specific skills in beginning tennis classes, and in normal sized classes such machines aid significantly in achievement in these skills.
2. If traditional techniques are to be rotated with such techniques reinforced by use of a machine such as the Ball-Boy, it is more effective to emphasize traditional techniques first and then add practice with the Ball-Boy.
3. It is possible that students, when speed and direction of practice balls are controlled, develop habits of performance which hamper performance in actual game conditions requiring a much larger variety of footwork patterns. The influence of such "sets" on game performance is recommended as a topic for further research (1965).

Printed and Static Material

Included under this heading are texts, workbooks, and mimeographed material, and such nonmoving visual aids as slides, transparencies, photographs, charts, and illustrations. These means of communication can orient the student, provide goal direction, and present actual content as well as mechanical principles that might effectively transfer to the learning of movements involved in skills.

Static visuals have usually been used to help students learn psychomotor skills. Written matter, besides serving this function, can be used to present rules and regulations, terms, equipment and space dimensions, and appropriate tactics. Many handbooks are available that describe and illustrate pertinent features of an activity; or you can write your own handouts.

Since little research has been done on the benefits of these kinds of media, it is difficult to make any generalizations about them. Their incorporation into the instructional process, singly or in combination, must be based on each teacher's analysis of his own circumstances. Less intricate visuals and printed material are necessary for younger students, beginners, and those with verbal difficulties. Printed and static visual materials are relatively cheap, are widely used in almost every class situation in one way or another, and can contribute to the student's learning.

Tape Recorders, Records, and Radios

Auditory media in the form of tape recorders, records, and radios generally rival or complement the teacher's oral communication, as well as the visual media already discussed, in orienting the student, gaining his attention, motivating him, presenting content, and administering instructions and directions. In addition, recordings encourage individualized approaches to learning.

Furthermore, recordings can cue responses, as in the learning of dance or other prescribed, systematic, sequential movements, on a group or individual basis. Also consider the value of tape recorders as a source of feedback. In large classes teachers do not have time to appraise each student's performance and behaviors. Or so much happens before the teacher and student can interact that the teacher might forget the key points he wished to communicate. The teacher could comment into the tape recorder as he observed the behaviors of each student and then students could play back the recording either in or out of class.

The lack of research literature on the value of these media in physical education does not permit any conclusions. In other areas, such as college teaching, military training, and radio communication courses, no differences in effectiveness have been noted between live lectures and tape recordings. The implications are that one can substitute for the other (at least in certain ways) and that these auditory media may have additional functions.

Automation and Programmed Materials

In recent years the major emphasis in education has been on the contribution of behavioral technology to completely automated education. Whereas most media have supplemented the teacher's efforts or partially replaced them, completely automated instructional systems are currently in operation in many aspects of education. The ultimate in automated instruction is the computer, a costly but versatile piece of equipment.

The best example of an instructional approach that does not require the teacher's presence is programmed instruction, which may appear in automated or nonautomated form. At first programmed instruction was presented by teaching machines, but now texts and packaged plans can deliver these programs too. There are various types of programmed instruction, but in general programmed methods may be compared with more traditional (teacher-centered) methods as follows:

". . . recordings can cue responses, as in the learning of dance or other prescribed, systematic, sequential movements, on a group or individual basis."

The traditional approach:

1. permits more creativity; the student is not bound to a predetermined learning pattern
2. allows more freedom to explore errors
3. is more adaptable to individual learning styles
4. allows interaction between the social and the learning environments

The programmed approach:

1. provides for students' individual learning rates
2. causes constant participation in the learning process, because students must respond in order to proceed
3. allows for constant and immediate feedback
4. can be used more effectively to teach large numbers of students with varying abilities
5. corrects errors before they persist
6. frees the teacher from paperwork and other routine chores, so that he can provide individual assistance

Until recently programmed instruction has been used almost exclusively in the cognitive domain, but it is based on sound theory and can be applied to the learning of any subject. Behavior is carefully shaped by means of feedback and reinforcement. Material is presented in small units, to which the student responds actively. He can readily obtain information as to the correctness of his response. If it was incorrect, he must go back and master the material; if correct, he is allowed to proceed to the next, more difficult unit. He acquires knowledge in an orderly fashion, and errors are discouraged.

Computers and teaching machines are currently being used and investigated in a variety of educational programs. In physical education programmed material has been developed to communicate information *about* skills, but material intended for the *practice* of skills was relatively nonexistent until recently. One of the major reasons for the lag is that programming skills practice is more difficult than programming cognitive matter. But so far, it appears that programmed guidance for both cognitive matter and psychomotor skills results in similar, if not better, results than traditional instructional methods.

Focus

In the article, "Prepackaged Sports Skills Instruction: A Review of Selected Research," Lawrence Locke and Mary Jensen analyze seven studies dealing with programmed instruction. The authors' analysis of the advantages and disadvantages of using programmed instruction in physical education provides penetrating insight into the issue of programmed versus traditional instruction. An excerpt from the article follows.

The theoretical benefits of programmed learning in the classroom are widely known. They include individualization, simplification, improved reinforcement, and prompt, accurate feedback. Presumably these advantages would obtain irrespective of the nature of the subject matter programmed. In theory, at least, a number of additional benefits might be derived from using programmed instruction in the physical education setting.

Since individuals proceed at their own learning rates, most students are assured not only some degree of success but also minimal mastery of material and skills which serve as the foundations for more advanced levels of skill. Mastery of fundamentals is particularly important for motor skills because of their progressive and cumulative nature; and the step-by-step presentation of programmed instruction gives the teacher and student some confidence that the learner really is ready for the next step.

Programmed instruction enables the teacher (even in the face of less than expert competence in the activity) to bring a well-structured, carefully thought out program of motor skill instruction to his students. This is particularly important in physical education because the field covers so many different sports, each with its own set of unique requirements.

The program, depending on its design, can control both the quantity and quality of practice for each student. In contrast, a teacher is faced with the choice of providing a limited amount of individualized instruction or of sacrificing individual instruction in favor of standardized instruction for everyone, such as occurs in mass calisthenics.

Programmed instruction assists the teacher in dealing with heterogeneity of student abilities by giving each student the material he needs when he is ready for it. The motor moron need not compete with the varsity athlete for the teacher's attention. With programmed texts, students can begin learning as soon as they arrive in class. There is no need to wait for other students to assemble or get ready since each learner is on a one-to-one relationship with the subject matter.

Each student can review skills already learned whenever he feels it's necessary, without holding back other students or interrupting a lecture or demonstration. Rules, history, etiquette, mechanical principles, and strategy could be programmed for each sport, thereby freeing the teacher from routine lectures and explanations and enabling him to spend more time with individuals who need supplemental help. In addition, take home programs for such material could be used to free even more class time for actual motor skill practice. (1971, p. 58)

The Research Problem

You may be confused because research has not demonstrated that, despite their extensive possibilities, media have had a major impact on instructional results in physical education classes. We will attempt to explain this contradiction. Prior experience, appeal to authority, and common sense, as well as scientific research, contribute to our knowledge of the world around us. Research is a valuable tool and is probably the most influential, concrete, and acceptable source of knowledge. However, one particular investigation, or even two or three, do not necessarily provide a sound basis for a conclusion that may have serious implications. When a number of acceptable studies consistently generate similar findings, a clear-cut basis for action exists. Research on the effectiveness of various forms of media in physical education leads us to question their worth. Let's look at some of the major drawbacks of much of the research conducted thus far.

First of all, very little research has been published on this subject compared with almost every other topic in physical education and with almost every other educational area, despite the recent interest of a number of graduate students, as revealed by thesis and dissertation titles. Until more research has been completed, therefore, conclusions about either a specific medium or media in general cannot be stated with any degree of certainty.

Secondly, designing real-life experiments, as compared with laboratory ones, is fraught with dangers and difficulties. Many variables are terribly difficult to control. In any educational field, studies comparing one mode of instruction with another usually reveal no differences in performance. Why?

It would be difficult here to explain the experimental design factors that contribute to this state of affairs, but we will mention some briefly. In order to show differences statistically, (a) large enough samples must be used, (b) groups should be reasonably similar at the start, (c) there should be minimum variability of evaluation scores, (d) evaluation techniques must be valid and sensitive, (f) enough time must be allocated to demonstrate learning differences, (g) there must be no teacher bias, and (h) confounding instructional and individual variables must be controlled. For various reasons these requirements have usually operated against the researcher. The failure of media researchers to consider these and other factors may help to explain why they have found media ineffective and also why one researcher might obtain data favoring one instructional method, and another researcher data favoring the other.

Perhaps a few cases will help to illustrate these points. A film may really have instructional value, but it is poorly done, has no soundtrack, or is not incorporated effectively into the instructional situation. VTR may really be able to reinforce acts and provide information feedback better than traditional methods, but the students may be too immature or not know how to analyze their acts on film. How much time is spent with media? How much in actual practice? In one study a special media group may be provided with the same amount of practice time as the comparison group, but with additional time in which to use the media. In another study both groups may be given the same amount of class time, but media usage may cut down on the available practice time for one of the groups.

Such dilemmas are endless. Designing media experiments and controlling potentially dangerous biasing variables is an extremely challenging task. Realizing these problems and the present state of research should help you understand why proponents of media and multi-media learning environments and those who favor more traditional teaching methods can argue legitimately for their own positions.

There is little doubt, however, that the wise deployment of various forms of media can aid individual learning. This feature may be good up to a point in physical education, in which social objectives are highly valued. Media can still be employed in socially dynamic situations, but they will probably differ and be used differently.

In concluding, it may appear that the purpose of this chapter was to sell media usage to physical educators. This is not really the intent. Media *should not* be used merely for the sake of using them. They are often costly, require knowledge of operational techniques, and need to be serviced. Once objectives have been stated and learning events sequenced, all possible instructional alternatives should be noted and reviewed. A particular form of media would ideally be chosen for its desirability and its superiority to others in fulfilling objectives. You will have to rely on sound judgment, learning theory, sparse physical education research, extensive educational and training research, and good luck.

Summary

We have examined the scope of the media available to physical educators and suggested ways in which they might be used. Many forms of media have proved to fare no better in performance results than traditional teacher roles, but media certainly help to enrich the instructional process in a number of ways. Instructional strategies determine the best possible presentation modes. Criteria for media selection were discussed, and various types were presented and analyzed, with practical suggestions for implementation. The instructional situation should be carefully analyzed before media are selected and used. Research findings and methodological weaknesses were also reviewed in order to explain the confusion about theory, research, and practice. At present it appears that when making decisions, you will have to rely more on learning theory, intuition, and educational and training research in general than on physical education research.

Wayne State University has a resource center on media in physical education under the direction of Dr. Chalmer G. Hixon. The center contains materials, information, videotapes, and audiotapes. According to publicity releases, the services available to members include duplication of videotapes and audiotapes for educational purposes; distribution of conference reports and bibliographies on the use of media in physical education; collection and storage of materials on the subject; master tapes of speeches, conferences, demonstrations, and instruction in physical education; and limited taping services for speeches, conferences, and demonstrations.

Resources

Books

Education

Briggs, Leslie J. *Handbook of Procedures for the Design of Instruction.* Pittsburgh: American Institutes for Research, 1970.   Systems approach to effective instruction.

Briggs, Leslie J., Campeau, Peggie L., Gagné, Robert M., and May, Mark A. *Instructional Media.* Pittsburgh: American Institutes for Research, 1967.   Clear and easy-to-follow writing on selecting media for instructional purposes.

Gagné, Robert M. *The Conditions of Learning.* New York: Holt, Rinehart and Winston, 1970, Chapter 12.   Good general guidelines on the use of media in instruction.

Gerlach, Vernon S., and Ely, Donald P. *Teaching and Media: A Systematic Approach.* Englewood Cliffs, N.J.: Prentice-Hall, 1971.   A number of chapters contain material useful in understanding and applying the contents of our Chapter 9.

Glaser, Robert (ed.). *Training Research and Education.* New York: Wiley, 1962.   Scholarly effort by noted experts on aspects of training, with special implications for education.

Knirk, Frederick G., and Childs, John W. (eds.). *Instructional Technology.* New York: Holt, Rinehart and Winston, 1968.   Book of readings covering various forms of media and mediums.

Lumsdaine, Arthur A. "Instruments and Media of Instruction." In N. Gage, (ed.), *Handbook of Research on Teaching.* Chicago: Rand McNally, 1963.   Good summary of research on various forms of media.

Pipe, Peter. *Practical Programming.* New York: Holt, Rinehart and Winston, 1966.   Functional book for writing materials associated with programmed instruction.

Torkelson, Gerald. *What Research Says to the Teacher: Instructional Media.* Washington, D.C.: Department of Audiovisual Instruction, National Education Association, 1969.   Summarizes and applies to teaching the research related to various kinds of media.

Physical Education

Biles, Fay R. (ed.). *Television: Production and Utilization in Physical Education.* Washington, D.C.: AAHPER, 1971.   Stated purpose is to acquaint physical educators with the process of instructional television usage.

Articles

Education

Tosti, Donald T., and Ball, John R. "A Behavioral Approach to Instructional Design and Media Selection." *AV Communication Review,* 17: 5–25, 1969.

Physical Education

James, P. E. "Video Feedback in Learning Beginning Trampoline." *Perceptual and Motor Skills,* 32:669–670, 1971.

Locke, Lawrence F., and Jensen, Mary. "Prepackaged Sports Skills Instruction: A Review of Selected Research." *Journal of Health, Physical Education, and Recreation,* 42:56–59, 1971.

Solley, William, and Borders, Susan. "Relative Effects of Two Methods of Teaching the Forehand Drive in Tennis." *Research Quarterly,* 36:120–122, 1965.

10

Applying Instructional
Materials and Procedures

In teaching and modifying behaviors certain factors must be considered—available media, equipment, and locales, as well as group size, homogeneity, and skill level. Decisions must be made about teacher-student roles and means of interaction and communication. The steps leading to the acquisition or development of each sample behavior in this chapter would be generally applicable to most kinds of objectives in that behavioral domain. Unfortunately, it is impossible to do justice to the numerous and varied objectives the physical education teacher might formulate for his class, but similarities in instructional techniques may be apparent enough to insure the value of this chapter.

Chapter Contents

Understanding This Chapter

The Instructional Sequence
Procedural Outline
Psychomotor Behaviors
Cognitive Behaviors
Affective Behaviors
Social Behaviors

Related Concerns
Availability of Equipment and Media
Learning Locale
Teacher-Student Roles
Group Size
Group Skill Level
Class Schedule

Summary

Student Objectives

The student who understands the material in this chapter should, with the book closed, be able to:

1. Identify and briefly explain the prelearning event (the GAMES plan) and the learning event (the PRACTICE plan) considera-

tions and presentation modes for the acquisition or development of

a. a self-paced psychomotor behavior
b. an externally paced psychomotor behavior
c. a cognitive behavior
d. an affective behavior
e. a social behavior

2. Describe six instructional considerations related to athletic equipment and media resources, learning locales, teacher-student roles, group size and skill level, and class schedule.

Understanding this Chapter

You may often be impressed with what you read in college but question its practical value. Examples are necessary to bridge the gap between theory and application, between the book and the gymnasium. The intent of this chapter is to show that the concepts presented in the previous chapters can in fact be applied to your teaching situation. Figure 10.1 illustrates that point in the systems model where we now find ourselves. Alternative presentation modes are stated for particular learning events. But by and large, the strategies are designed and presented to minimize confusion and so will appear to be prescriptive, although they are not intended to be; they are not the only ones acceptable for the desired conditions.

Performance objectives will be stated for each of the four major types of behaviors. The overall strategy for the learning of a

FIG. 10-1 *Systems approach model*

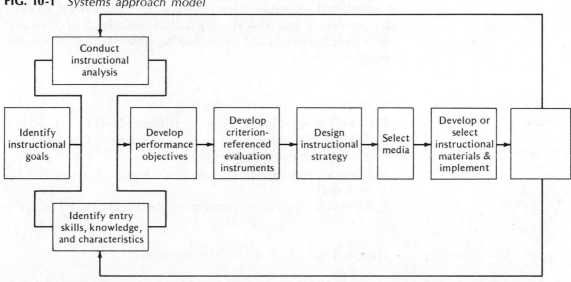

psychomotor task is slightly different for self-paced and externally paced activities. Examples of both types are presented. Strategies also vary for cognitive behaviors, depending on whether they involve recall, comprehension, analysis, problem solving, or decision making. Only one example is offered, however; it represents a combination of recall and decision making. Likewise, there are different strategies for affective behaviors associated with motivation and interest or attitudes and values. The latter category is illustrated on the following pages. Finally, one type of social behavior, interpersonal relations, is presented, although there are others, such as self-fulfillment, emotional stability, and conduct. Bear in mind that similar strategies may be used for different behaviors, but specific instructional considerations are related to the particular type of behavioral classification within each domain.

We will formulate acceptable strategies step by step, sequencing learning events and suggesting presentation modes. We suggest that you try to sense the mood of the group and individual exceptions when determining what steps, if any, should be taken with regard to the prepractice or preexperience events. You will not have to direct these events if the student undertakes them on his own initiative. Since every teaching situation is different and it would be impossible to consider every possible variable, certain basic assumptions are made at the start (but variations might dictate modifications in strategies), i.e., that the physical education class is of average size (about thirty to forty-five students), that it takes place in a high school, that media, facilities, equipment, and space are adequate, and that the students have had no experience with the stated performance objectives. Let us also assume that the instructional goals have been identified, the instructional analysis has been made, and student entry skills, knowledge, and characteristics have been evaluated. The following sequence suggests how we might proceed.

The Instructional Sequence

Procedural Outline

1. Write the objective (Chapter Six)
2. Classify it according to behavior (Chapter Five)
3. Check to make sure your classification is correct (Chapter Five)
4. Analyze the task (Chapter Five)
5. Prepare the student for learning (the GAMES approach) (Chapter Eight)
6. Select and sequence appropriate learning events (the PRACTICE approach) (Chapter Eight)

Psychomotor Behaviors

Example One: The Crawl Swimming Stroke

1. The objective is to have the students:

   perform the crawl swimming stroke according to American Red Cross standards.

2. The objective is classified as:

   psychomotor

3. Is the behavior primarily movement oriented? Is it concerned with:

   manipulation (contact with objects, such as grasping, controlling, releasing)

   locomotion (movement of the body in space)

   propulsion (catching or throwing objects)

   control (body balance, balancing objects)

   Yes, the objective is classified as psychomotor because it involves locomotion, control, and continuous coordination of the body parts.

4. Is the task self-paced or externally paced?

   It is self-paced:

   the student initiates and executes movements at his own speed

   the situation is stable and predictable

5. Preparing the student (the GAMES approach):

Prepractice Events

Goal orientation

   The student should be aware of American Red Cross standards for the crawl stroke; the image of the ideal stroke should be reasonably clear in his mind. He should know what is expected of him.

Attention

   The student should be alert and responsive to instruction and appropriate cues.

Presentation Modes

visual, verbal

   Observation of performance (live or on film), illustrations and photographs, and explanations.

visual, verbal, kinesthetic

   The learning environment—air and pool temperatures, lighting, teaching techniques, atmosphere, etc.—should stimulate and maintain attention

". . . the objective is classified as psychomotor because it involves locomotion, control, and continuous coordination of the body parts."

Motivation

The student should be optimally motivated to perform.

verbal

Since the crawl is moderately difficult for most students, general incentives should probably be moderate also. Externally and internally oriented motivation may be applied where appropriate, although most students should be sufficiently motivated by the challenge of accomplishing this skill.

Experienced for transfer

The student should have experienced prerequisite skills and knowledges that he can transfer in a positive manner to the present situation. Furthermore, he should be informed of these relationships.

verbal, visual

Assuming the presence of prerequisite skills and knowledges, verbal explanations or observations should emphasize and clarify their relationship to the present task.

Situation familiarity

The student should be familiar with the essence of the learning situation, the nature of the general cues to which he will respond (in this case, the medium of water and the task of propulsion in a prescribed manner).

visual, verbal

Discussion of water and demonstration of effective propulsion in terms of physics and mechanics.

6. Selecting and sequencing appropriate learning events (the PRACTICE approach):

Practice Events

Practice

The student should practice often and take advantage of self-paced task benefits: being able to utilize cues, feedback, and reinforcement while continually executing the act.

Presentation Modes

kinesthetic, visual, verbal

Performing the task in the pool or at poolside, conceptualizing the task. Since the task is self-paced, the environment is relatively stable and the student need not be concerned with sudden appearing, unpredictable stimuli to which he must respond instantaneously. Emphasis is on response, and the primary goal is consistency of desired movements. Overpractice for long-term retention benefits.

Reinforcement

The student should be at least intermittently reinforced as he approaches skill mastery.

verbal

Comments, praise, rewards, grades, other forms of acknowledgment.

Administration

At least at first, practice sessions should be reasonably spaced and specific actual task practice reasonably distributed.

kinesthetic, verbal, visual

Fatigue, danger, and the learning of incorrect responses can occur during massed practice. Actual practice periods can be interspersed with rest or other forms of practice, as well as observing filmed or live performances.

Contiguity

The student should execute with appropriate timing the series of stimulus-response acts that constitute the chain of events of which the crawl stroke is composed.

kinesthetic, verbal

Individual components of the task are cued by the teacher and practiced in proper spatial and temporal arrangement.

Tradeoffs

The student should practice parts of the task and then the entire task; and he should stress accuracy of movement instead of speed (not swimming in competition!); physical instead of mental practice; social (paired) instead of individual practice; and real instead of simulated conditions in most cases.

kinesthetic

Correct execution is more important than speed; task complexity and organization dictate use of the part-method approach (a) moving legs together in flutter kick, (b) breathing, (c) moving arms, (d) integrating entire stroke); practice is primarily physical, but the student should visualize the correct routines and think about applicable move-

kinesthetic (continued)

ment and buoyancy principles after some actual practice, prior to further practice, and intermittently thereafter; paired practice should be emphasized to facilitate social interaction—students alternate roles of teacher and performer for guidance, feedback, and reinforcement; and simulation (deck practice, kick boards) can be used—infrequently—where appropriate to emphasize cues and correct responses.

Information feedback

The student should be aware of his skill development as he proceeds in practice.

kinesthetic, verbal, visual

Internal (kinesthetic) sources provide some concurrent information about form and execution. Teacher comments and videotape are external sources of terminal feedback and more precise information.

Cue generalization and discrimination

The student should proceed from general concepts and responses to highly specific acts.

kinesthetic, verbal, visual

Meaningful and appropriate cues at each phase of learning (words cue and direct the student during the sequence of movements, large mirrors at poolside allow him to view body motions and body-part relationships, manual manipulation of passive limb(s) gives him the feel of the movement, model films and videotape continually inform him where he is and where he should be). Concentration and attention are not overly important during practice of this task, as it is repetitious, involves large muscles of the body, and is self-paced.

Event repetition

The student should repeat the task frequently according to specific requirements.

kinesthetic

In the pool, assuming that is where the student will be tested.

Alternatives: practice conditions may vary as to depth (from shallow to deep water in the pool) and type of water (pool, lake, river, ocean). Although the swimming stroke remains relatively constant regardless of the environment, water type and depth may concern students. A variety of swimming experiences would be desirable, although they are probably not practical during the allotted time.

Example Two: Dance Movements

1. The objective is to have the students:

   execute a series of three leaping patterns in good form and synchronized with those of a partner.

2. The objective is classified as:

   psychomotor

3. Is the behavior primarily movement oriented? Is it concerned with:

   manipulation (contact with objects, such as grasping, controlling, releasing)

   locomotion (movement of the body in space)

   propulsion (catching or throwing objects)

   control (body balance, balancing objects)

   Yes, the objective is classified as psychomotor, because it involves locomotion, control, and continuous coordination of the body parts.

4. Is the task self-paced or externally paced?

   It is externally paced:

   the student initiates and executes movements according to pre-designated cues and his partner's movements.

5. Preparing the student (the GAMES approach).

Prepractice Events

Goal orientation

The student should know the individual dance movements as well as the synchronized patterns of energy expected of him.

Attention

The student should be alert and responsive to instruction and appropriate cues.

Presentation Modes

visual, verbal

Observation of performance (live or on film), illustrations and photographs, and explanations.

visual, verbal, auditory

The learning environment atmosphere — teacher techniques, music choice, choreography, etc. — should stimulate and maintain attention.

Motivation

The student should be optimally motivated to perform.

verbal, visual, auditory

The task is moderately difficult, and general incentives to stimulate learning should probably be moderate also.

Experienced for transfer

The student should have experienced prerequisite skills and knowledges that he can transfer in a positive manner to the present situation. Furthermore, he should be informed of these relationships.

verbal, visual

Assuming the presence of prerequisite skills (leaping ability, timing of sequence) and knowledge, verbal explanations or observations should emphasize their relationship to the present task.

Situation familiarity

The student should be familiar with the essence of the learning situation, the nature of the general cues to which he will respond. In this case, he will perform in a special gymnasium, classroom, or studio, in conjunction with musical tempos or sounds (rhythmics) and another moving body.

visual, verbal

Demonstration, discussion of elementary dance movements and the principles of synchronization.

6. Selecting appropriate learning events (the PRACTICE approach) and sequence:

Practice Events

Practice

The student should practice as frequently as necessary until he has mastered the task. Since it is externally paced, although the cues occur predictably, it is only moderately difficult. The correct timing of responses should be emphasized.

Presentation Modes

kinesthetic, visual, verbal

Performing the task, first without, then with a partner. Observing and thinking about the activity, and listening to and perceiving tempo sounds result in consistent movements. Overpractice to resist stress and fatigue interference.

Reinforcement

The student should experience intermittent reinforcement as he approaches mastery.

verbal

Comments, praise, rewards, grades, other forms of acknowledgment.

Administration

The student should master this task, with the necessary prerequisite skills, in one or two practice sessions. Specific actual task practice should be relatively massed unless the student is apparently not benefiting after reasonable time and effort.

kinesthetic, verbal, visual

The task should not involve many competing responses, and skill should be demonstrated before fatigue sets in. If skill is not being attained, rest or other forms of learning may be appropriate.

Contiguity

The student should execute with appropriate timing the series of stimulus-response acts

kinesthetic, verbal

Individual components of the task are cued by the teacher and auditory tempos, prac-

Contiguity (continued)

that constitute the chain of events of which synchronized leaping movements are composed.

Tradeoffs

The student should practice parts (individual leaps) and then the entire task, at the desired speed in correct form; practice should be primarily physical instead of mental and should occur under real (paired practice) instead of simulated conditions.

Information feedback

The student should be aware of his skill development as he proceeds in practice.

Cue generalization and discrimination

The student should proceed from general concepts and responses to highly specific acts.

Event repetition

The student should repeat the task frequently according to specific requirements.

kinesthetic, verbal (continued)

ticed in proper spatial and temporal arrangement. Timing is extremely important in this task.

kinesthetic

Equal emphasis on speed and accuracy of movement; slight if any emphasis on part practice (depends on extent to which prerequisite skills are developed); primarily physical practice, but the student should imagine the correct cues, routines, and applicable movement principles, and think intermittently about the timing of the movements throughout practice regimens; and simulated practice should be unnecessary and impossible to develop effectively.

kinesthetic, verbal, visual

Internal (kinesthetic and visual) sources provide some concurrent information about form, execution, and timing. Teacher comments and videotape are external sources of terminal feedback and more precise information.

kinesthetic, visual, verbal, auditory

Meaningful and appropriate cues at each phase of learning (words, tempos, and partner's movements influence the student during the sequence of movements; large mirrors allow him to view body motions; model films and videotape continually inform him where he is and where he should be). Concentration and attention are fairly important during practice as the task involves precisely executed cued responses. This is especially true if no external tempo is present.

kinesthetic

On the dance floor, or wherever the student will be tested.

Alternatives: students may practice different directions of movements; for example, partners moving toward each other, or at diagonals. Other variations might include going into the leaps from different movement posi-

"The student should be aware of his skill development as he proceeds in practice."

kinesthetic (continued)

tions, such as following a turn or while in a turn. Student might also perform with music or sounds in tempo or without external cues, in which case internal rhythm must be kept.

Cognitive Behaviors

Example: Volleyball Rules

1. The objective is to have the students:

   observe a volleyball game and be able to identify a legal setup, serve, dig, spike, block at the net, net violations, and legal player rotations.

2. The objective is classified as:

   cognitive

3. Does the behavior primarily involve knowing? Is it concerned with:

   recall (remembering facts, ideas, procedures)

   comprehension (interpreting, translating, extrapolating)

   analysis (organizing patterns, relationships)

   problem solving (applying ideas, evaluating)

   decision making (selecting, classifying)

   Yes, the objective is classified as cognitive because it involves recall of facts and ideas, analysis, and decision making.

4. Is the task one of recall, comprehension, analysis, or decision making? It is one of recall and decision making:

the student remembers facts and ideas

the student analyzes and classifies behavioral occurrences

5. Preparing the student (the GAMES approach):

Preexperience Events

Goal orientation

The student should be informed of all the rules of volleyball he should learn, to what degree, and the context and nature of the testing situation.

Attention

The student should be prepared to respond to instructions, reading materials, and play situations.

Motivation

The student should be optimally motivated to perform.

Experienced for transfer

The student should have experienced prerequisite skills and knowledges that he can transfer in a positive manner to the present situation. Furthermore, he should be informed of these relationships.

Situation familiarity

The student should be familiar with the learning situation and the nature of the testing procedures (where tested, how tested, how much tested for).

Presentation Modes

verbal, visual

Explanations, reading materials, observations of and reactions to play situations.

verbal, visual

High-quality learning environment and learning materials.

verbal, visual

Need for decision making in limited time suggests a moderate degree of motivation appropriate for that task. Overly strong incentives may tend to impair cognitive performance; weak ones may not stimulate students' thought processes and help them make correct decisions.

verbal, visual

Explanation of relationship and necessity of rules from sport to sport. Previous experiences in memorizing facts and ideas, then applying them to, analyzing, and classifying presented materials, should be related to the volleyball experience, which deals with identifying and classifying behaviors.

visual, verbal

Providing information, demonstration, observation.

6. Selecting and sequencing appropriate learning events (the PRACTICE approach):

Experience Events	*Presentation Modes*
Experience (practice)	verbal, visual
The student should have sufficient time and the most favorable conditions for learning. A rule book and the game situation, in which analyses and decisions about rule infractions must be made, should supplement each other.	Reading about and explanation of rules, observation of infractions, with students actively participating in decision making on occasion.
Reinforcement	verbal
The student should experience intermittent reinforcement as he begins to master the information and make correct decisions.	Comments, praise, rewards, grades, and other forms of acknowledgment as the student displays his knowledge (through informal tests, calls made on himself during play).
Administration	visual, verbal
The student's experiences should be distributed (reading assignments and decision calls made in play or while observing play).	Frequent, spaced experiences should result in desired behaviors.
Contiguity	visual, verbal
The student should make immediate decisions about infractions.	Frequent reading, observations of right and wrong playing techniques. Students should respond immediately to situations, as merely being informed does not insure the development of the identification and decision-making processes.
Tradeoffs	visual, verbal
The student learns rules as parts, then responds to the total game situation, and under real rather than simulated conditions.	Students first master individual rules and infractions, then analyze the game in progress or particular behaviors for infractions. Real conditions are favored but situations can be contrived to emphasize certain points and reinforce the acquisition and application of knowledge.
Information feedback	verbal
As he progresses, the student should be aware of his information acquisition as well as his ability to apply it in making decisions.	Self-testing, formal and informal teacher questions, observations of students' responses to infractions in play.
Cue generalization and discrimination	visual, verbal
The student should proceed from general concepts of correct play to highly specific interpretations of and decisions on specific acts.	Teacher prompts, models decisions.

Event repetition

The student should frequently study rules and then make decisions while observing play or in play when appropriate.

visual, verbal

Reading appropriate matter, experiences in gymnasium.

Alternatives: Students may experience a wide range of behavioral infractions to enable them to recognize violations wherever and whenever they occur. Particular rule concepts or infractions can be presented in more than one way. Providing a choice of media and learning resources may help students who learn better with one presentation mode than another.

Affective Behaviors

Example: Attitudes Toward Physical Fitness

1. The objective is to have the students:

realize the body can move more effectively and that the possibility of ill health decreases as physical fitness increases. Students will demonstrate this realization by spending more free time in physical fitness-oriented activities after completing the course.

2. The objective is classified as:

affective

3. Does the behavior primarily involve feeling? Is it concerned with:

evaluation (selection, acceptance)

motivation (interest, commitment)

Yes, the objective is classified as affective because it involves valuing, accepting, and stating a preference.

4. Is the behavior indicative of motivation and interest or attitude and value?

It is primarily attitudinal:

the student chooses an activity voluntarily without being pressured or observed.

5. Preparing the student (the GAMES approach):

Preexperience Events	*Presentation Modes*
Goal orientation	verbal, visual
The student should be informed of the goal but not of the specific objective (acts). It is inadvisable for the teacher to inform students how they will be evaluated; otherwise, they may deceive him.	Verbalized attitude, determine present feelings, and illustrate values of physical fitness.
Attention	verbal, visual
The student should be alert and responsive to instruction and situational cues.	High-quality learning environment to stimulate and maintain attention (atmosphere, teacher techniques).
Motivation	verbal, visual
The student should be optimally motivated to modify existing attitudes or maintain present desirable ones.	Illustrations, observations, reading matter, and verbalization to show the importance of acquiring the attitude.
Experienced for transfer	verbal
The student should have experienced situations that have fostered positive attitudes toward participation in a variety of physical education and fitness activities, social activities, intellectual activities, and the like.	Explanation of the relationship between other activities vital to man's well-being and the present circumstance.
Situation familiarity	visual, verbal
The student should be familiar with the physical fitness program and environment in which he will be expected to develop favorable attitudes toward physical activities.	Observing, informing, discussing.

6. Selecting and sequencing appropriate learning events (the PRACTICE approach):

Experience Events	*Presentation Modes*
Experience (practice)	verbal, visual, kinesthetic
High-quality experiences over a sufficient period of time should enable the student to learn the importance of physical fitness and relationships to total well-being.	Reading materials, participation in activities, discussions. Carefully designed experiences will help the student critically analyze current attitudes and modify them accordingly. Attitudes are usually not changed easily, and negative ones must be handled delicately by showing the values to be derived from positive attitudes and resultant activities. Understanding and cognition are a part of attitudinal change.

"High-quality experiences . . . should enable the student to learn the importance of physical fitness and relationships to total well-being."

Reinforcement

The student should receive intermittent reinforcement as he demonstrates knowledges and practices associated with positive attitudes toward physical fitness activities.

verbal

Comments, praise, rewards, and other forms of acknowledgment as students show positive attitudes. Patience is necessary; changing negative attitudes is often a slow and tedious process.

Administration

The student should experience distributed social and individual situations geared toward the realization of favorable attitudes.

verbal, kinesthetic

Frequent, spaced, and varied experiences.

Contiguity

The student's desirable responses should be reinforced immediately.

verbal

Teacher sensitivity to "true" behaviors reflecting favorable attitudinal changes.

Tradeoffs

The student should experience a variety of situations, not just one; free choice instead of prescribed ones, and real instead of artificial situations.

verbal, visual, kinesthetic

Variety of experiences takes into account differences in students' responsiveness and the fact that one experience reinforces another; free choices are not restrictive or biasing and tend to lead to true changes; role playing or simulation can effectively demonstrate negative or positive outcomes of attitudes, but experience in the real situation in which the attitude is to be adopted is more desirable.

Information feedback

The student should be aware of his progress as his attitudes favorably change.

visual, verbal

Informal observations and records of in-class and out-of-class behaviors, discussion of implications.

Cue generalization and discrimination

The student's general, unclear attitudes about the nature and benefits of physical activity should be replaced by precise realizations of effects.

verbal, visual, kinesthetic

A variety of experiences, teacher prompts, and guidance.

Event repetition

The student should frequently experience a variety of situations intended to shape positive attitudes toward physical fitness.

visual, verbal, kinesthetic

Reading pertinent materials; psychomotor, cognitive, and affective experiences in class.

Social Behaviors

Example: Interpersonal Relations

1. The objective is to have the students:

   display appropriate competitive and cooperative behaviors when participating in basketball. This goal will be evaluated through a check list of observations, and its achievement demonstrated by more desirable behaviors in later rather than earlier experiences in the activity.

2. The objective is classified as:

   social

3. Is the behavior primarily oriented toward personal and social adjustment? Is it concerned with:

   conduct (sportsmanship, honesty, respect for authority)

   emotional stability (control, maturity)

   interpersonal relations (cooperation, competition)

   self-fulfillment (confidence, self-actualization, self-image)

   Yes, the objective is classified as social because it involves cooperative and competitive behaviors.

4. Is the behavior related to personal adjustment or social adjustment?

   It involves social adjustment:

   the student learns to compete against others in a socially acceptable manner and to cooperate with teammates.

5. Preparing the student (the GAMES approach):

Preexperience Events	*Presentation Modes*
Goal orientation	verbal
As in the case of affective behaviors, the student should be informed of the goal but not of the specific objective, and should not be told how he will be evaluated. True competitive and cooperative inclinations can be assessed only if the student is unaware of how he is being appraised and if he is not made to conform to expectations.	Verbalize the interpersonal relations goal, determine present tendencies, and explain rationale behind the development of favorable interpersonal relations.
Attention	verbal, visual
The student should be alert and responsive to instruction and situational cues.	High-quality learning environment, geared to stimulate and maintain attention (atmosphere, teacher techniques).
Motivation	verbal, visual
The student should be optimally motivated to modify existing cooperative and competitive tendencies or maintain present desirable ones.	Explanations and observation of effects of desirable behaviors on group will show the importance of developing ideal interpersonal relations.
Experienced for transfer	verbal
The student should have experienced a variety of situations that have tended to encourage cooperation and competition.	Explanation of relationship between cooperative and competitive behaviors in other situations and in the present one. Explanation of why these behaviors are necessary in many forms of organized activity.
Situation familiarity	verbal, visual
The student should be familiar with the game of basketball and the court, in and on which he will be expected to display appropriate competitive and cooperative behaviors.	Informing, discussing, observing.

6. Selecting and sequencing appropriate learning events (the PRACTICE approach):

Experience Events	*Presentation Modes*
Experience (practice)	verbal, visual, kinesthetic (continued)
High-quality experiences over a sufficient period of time should enable the student to learn the importance of desirable competitive and cooperative behaviors in sport.	Participation in basketball, discussions. Experiences should demonstrate the value of cooperative and competitive behaviors for the sake of continuity of the game and the group's enjoyment of it, as well as for individual fulfillment. These behaviors are slow

verbal, visual, kinesthetic (continued)

to develop, especially if the student initially demonstrates undesirable interpersonal relations.

Reinforcement

The student should receive intermittent reinforcement as he demonstrates more desirable cooperative and competitive behaviors in basketball playing situations.

verbal

Comments, praise, rewards, and other forms of acknowledgment as the student develops appropriate behaviors. Patience is necessary.

Administration

The student should experience distributed experiences during the unit designed to promote behaviors.

verbal, kinesthetic

Frequent, spaced, and varied experiences.

Contiguity

The student should be immediately reinforced for desirable behaviors.

verbal

Teacher sensitivity to "true" behaviors reflecting favorable changes.

Tradeoffs

The students should experience a variety of situations, not just one; free choice instead of prescribed ones; and real instead of simulated situations.

verbal, visual, kinesthetic

A variety of experiences takes into account differences in students' responsiveness and the fact that one experience reinforces another; free choices are not restrictive or biasing and tend to lead to true changes; role playing or simulation can effectively demonstrate negative or positive outcomes of cooperative and competitive behaviors, but experience in the real situation in which the behaviors are to be expressed is more desirable.

Cue generalization and discrimination

The student's general understanding of the value of desirable cooperative and competitive behaviors should become specific realizations of their importance during basketball contests.

verbal, visual, kinesthetic

A variety of experiences, teacher prompts, and guidance.

Event repetition

The student should frequently experience a variety of situations intended to shape appropriate competitive and cooperative behaviors.

visual, verbal, kinesthetic

Participating in basketball games under various circumstances.

Related Concerns

Instruction cannot proceed effectively unless other matters are considered. For instance, what content is to be mastered? What athletic equipment and media resources should be available in the teaching situation, and which ones actually are? Where will learning occur—in a gymnasium, pool, classroom, athletic field, or elsewhere? What role or roles should the teacher and the students assume? How many students are there, and what is their general skill level? The answers to these and other important questions will help to influence instructional strategies. Since these factors tend to vary from one teaching situation to another the teacher should modify his strategies accordingly.

Availability of Equipment and Media

Instructional strategies reflect the availability of appropriate athletic equipment and media, and it should go without saying that teaching would be much easier if all the physical resources we could want were obtainable on request. This is an unlikely circumstance, but a creative and diligent teacher can develop makeshift materials, borrow from other sources, and modify teaching procedures within the limits of the situation. Often an instructional technologist, human resource specialist, audio-visual expert, media specialist, or whatever he may be called, is available to assist teachers in designing and improving instruction. Advice and media loans are his potential contribution.

Learning Locale

Physical education experiences need not only be associated with the gymnasium and other apparently appropriate locations. Field trips, visits to various locales within the school, and experiences in different settings can enable students to achieve outcomes they probably could not attain in the traditional classroom. Special preparations are in order when the class is to meet in a different place. Assumedly the alternative will fulfill certain enabling objectives that the traditional learning locale cannot. A class in softball may learn skills and strategies better when viewing an actual league softball game than in practice. They may develop an appreciation for the art of skilled movement. The overriding consideration in choosing the locale for any physical education activity is where and how enabling objectives can *best* be achieved.

Remember, too, that changed environments may provide a Hawthorne effect, i.e., improved motivation and attitude leading to better output, which is said to result when students do not continually experience routine and consequently boring teaching procedures in the same setting. Named after the desirable effect noticed on workers at the Hawthorne works of the Western Electric Company in a series of studies undertaken from 1927 to 1939, the Hawthorne effect has served as the basis for much speculation about the nature of working and learning environments. In the studies practice routines, working environments, and other factors were deliberately manipulated to determine their effects on performance.

Production seemed to be helped by the changes, but the fact that the workers knew they were participating in a study was perhaps even more important. In any case, the conclusion seemed to be that changes in scenery can be beneficial.

Teacher-Student Roles

If we agree that the teacher manages the learning process and is not the sole source of knowledge, then the ways in which he can interact with students must be identified. The student can be directed toward objectives through a variety of activities. One of the teacher's roles is that of *lecturer*, with the students memorizing information and performing as they are told. This technique is quite popular and serves definite purposes. However, it is not the only way to proceed.

If the teacher is trying to encourage such cognitive processes as analysis, evaluation, flexible thinking, higher-order organization of materials, probing, inquiry, and problem solving, other techniques may be advantageous. For instance, *discussion* encourages the exchange of ideas. With the teacher as discussion leader, the group ultimately resolves an issue or problem. Or the teacher may create a *situation* that will offer certain experiences to the student. The student may have to work out a problem by becoming personally involved in it. Discussions and experiences tend to encourage active involvement, whereas in the lecture method the student is relatively passive.

Students may *role-play* in order to become more sensitive to particular problems. For instance, the role of officials in various sporting events is not well understood. Students could be asked to take turns assuming officials' roles. In order to teach sportsmanship and honesty in sport, a student could play the role of antagonist, an official or an opponent who deliberately makes bad calls and violates rules. A good way for a group of students to learn the value of abiding by rules would be to see how such behavior disrupts a game. Don't forget that every teacher has a particular teaching style, which may encourage the use of one technique over another.

Group Size

The instructional procedures and individual considerations used in teaching an individual student do not apply with a group of thirty students. How about a class of fifty? In large cities classes may contain 100 to 200 students. A possible solution is the formation of subgroups. Groups of various sizes, within the context of available facilities and equipment, will place unique demands on the teacher. The larger the group, the greater the challenge to the teacher to mediate effectively the learning processes of individual students.

Group Skill Level

In many respects it is easier for the teacher to deal with homogeneous groups. Beginners in an activity are not treated like advanced learners, at least with regard to the emphasis placed on selected instructional events. For most favorable results, instructional strategies must coincide with the students' skill level. When the class is quite heterogeneous in terms of skill, the teacher usually tries to estimate the highest goal for most of the students. Although he cannot easily provide individual instruction for students at either extreme, various forms of media and certain types of instruction make it possible to do so. The group can be reorganized into subgroups according to capabilities, and techniques appropriate to these groups or individuals should be utilized.

Poorly skilled students are more dependent on external sources of information to guide their actions than skilled students. With skill, movements become more internalized. When a low level of skill is due merely to lack of experience, then traditional instructional methods should be used. Modeling of expected behaviors is important—objectives should be clearly understood. The teacher should emphasize the most important cues during the various stages of skill development since, if left to his own resources, the learner will probably attend to too many irrelevant stimuli at one time. More skilled students show an ability to filter out unimportant stimuli, to respond on occasion as if under automatic control, with less need for external prompting.

When students remain at a low skill level for a while, a careful analysis and diagnosis are in order. Poor motivation and attitude or poor grasp of prerequisite skills for the present learning tasks may be responsible. Another cause may be a presentation of instructional events which is incompatible with the student's preferred style of learning. Specially designed experiences should be prepared. Identification of the learning problems should indicate the measures necessary to remedy them. Specially designed media including simulation, redemonstrations of the right and wrong ways of executing the skill, manual manipulation, and emphasis on accuracy instead of speed may all prove to be beneficial when previous methods have failed.

On occasion, there is justification for heterogeneous groupings. Slow learners can profit from interaction with more skilled performers; nothing is more frustrating for beginners in tennis than opposing each other. One beginner rarely provides the other with enough opportunities to achieve the necessary skills. The highly skilled performer can assume leadership roles, assist the teacher, and help beginners develop. Arguments for and against homogeneous or heterogeneous groupings are numerous, but either method can help the teacher reach different goals.

Class Schedule

In most teaching situations, class schedules are dictated by the administration. In more progressive schools, they are more flexible. In either case the time allocated per class period and the number of periods per week will influence the kinds and numbers of objectives that can be realistically expressed. These factors will also affect the type and implementation of the learning strategies selected for class use.

For example, shorter or longer class meetings will help you determine which media, if any, to select. They define the amount of time that can be spent for each learning event, in any medium, and for experiencing the strategies geared to help the student achieve any instructional objective. Longer classes and/or classes that meet more frequently give the teacher greater expectations of fulfilling more and higher-level objectives.

Summary

In this chapter you have had an opportunity to observe, from the statement of the performance objective to the presentation of learning events and presentation modes, the instructional decisions for various behaviors. Sample strategies for psychomotor, cognitive, affective, and social behaviors were presented, and although they by no means exhausted all possibilities, hopefully they conveyed the ideas expressed in the previous chapters in a clearer, simpler, and more directly applicable format. At this point you should be able to develop appropriate learning events and presentation modes for any performance objective related to physical education. Any strategy, however, is contingent on such considerations as available athletic equipment and media, learning locales, group size and skill level, and class schedules.

Resources

Books

Education

Briggs, Leslie J. *Sequencing of Instruction in Relation to Hierarchies of Competence.* Pittsburgh: American Institutes for Research, 1968. Technical but informative review of research literature and interpretations for instructional decisions.

Gagné, Robert M. *The Conditions of Learning.* New York: Holt, Rinehart and Winston, 1970. Presents types of learning and learning events. May be a classic in its time.

Gerlach, Vernon S., and Ely, Donald P. *Teaching and Media: A Systematic Approach.* Englewood Cliffs, N.J.: Prentice-Hall, 1971. A number of chapters contain material useful in understanding and applying the contents of our Chapter 10.

Haney, John B., and Ullmer, Eldon J. *Educational Media and the Teacher.* Dubuque, Iowa: Brown, 1970. Practical approaches to shaping learning.

Merrill, M. David. "Psychomotor and Memorization Behavior." In M. David Merrill (ed.), *Instructional Design: Readings.* Englewood Cliffs, N.J.: Prentice-Hall, 1971. Easy-to-follow explanation of relation between psychomotor behaviors and learning conditions and processes.

Merrill, M. David, and Goodman, R. Irwin. *Instructional Strategies and Media: A Place to Begin.* Monmouth, Ore.: Teaching Research Division, Oregon State System of Higher Education, 1971. Although in rough form, contains good ideas and techniques for improving psychomotor, cognitive, and affective behaviors.

Physical Education

AAHPER. *Organizational Patterns for Instruction in Physical Education.* Washington, D.C.: AAHPER, 1971. Practical and innovative ideas and organizational patterns for organizing time allotment, personnel, and student grouping structure in physical education.

Mosston, Muska. *Teaching Physical Education.* Columbus, Ohio: Merrill, 1966. Alternative strategies for reaching goals, and recommended ones with good examples.

Singer, Robert N. *Coaching, Athletics, and Psychology.* New York: McGraw-Hill, 1972. Analysis of learning research and theory as applied to coaching and athletics.

Singer, Robert N. *Motor Learning and Human Performance.* New York: Macmillan, 1968. Learning research and theory as related to motor skills of all kinds.

Articles

Education

Tosti, Donald T., and Ball, John R. "A Behavioral Approach to Instructional Design and Media Selection." *AV Communication Review,* 17:5–25, 1969. Creative approach to instruction and presentation modes.

Physical Education

Dougherty, Neil J. "A Plan for the Analysis of Teacher–Pupil Interaction in Physical Education Classes. *Quest,* Monograph XV:39–50, 1971. Discusses the nature of verbal interaction between teacher and students. Other excellent articles are also found in this issue, e.g. "Educational Change in the Teaching of Physical Education."

Gentile, A. M. "A Working Model of Skill Acquisition with Application to Teaching." *Quest,* Monograph XVII:3–23, 1972. Applies knowledge of the process of motor learning to the teaching of skills. Other articles in this issue, e.g. "Learning Models and the Acquisition of Motor Skill," are also worthwhile.

11

The Revision Process

Formative evaluation and revision of an instructional activity is the final step in the systems approach. However, it is misleading to talk about a "final" step, as if the model were simply a sequence of activities to be completed. The model is, in effect, a circular one; i.e., based upon the data from a single implementation of an instructional activity, the teacher revises the activity and reimplements it.

The process is circular because it is extremely difficult to design effective instructional activities on the first attempt. If the teacher tests out his instructional design with one or two students prior to implementing it with a large group, he is sure to identify some of his major problems. However, he will have to conduct a major field trial with a large number of students before he can identify the majority of the problems. Therefore, the data from this field trial will be a starting point for revising all the work that has preceded it. The teacher begins to revise in order to increase the probability of reaching his instructional goal.

Numerous educators have argued that teaching is an art, that it depends on the teacher's personality, and that his interaction with students causes them to learn. The systems approach to instruction provides the basis for a more scientific attitude. While the systems approach emphasizes the analysis of behavior and the collection of data on student performance, there is still a great amount of "art" associated with it. Nowhere is this more obvious than when the teacher reaches the revision stage of instruction. The proponents of the systems approach have not defined the revision process very systematically; most have simply indicated that data should be collected and that revision should take place. Just how the data are used and what is revised has been left to the individual teacher. In this chapter we will suggest a number of procedures he can use in summarizing data in order to determine what types of revision are necessary. However, you should recognize that the revision process is still quite undeveloped, and a great deal of independent judgment must still enter into it.

Chapter Contents

Summarizing Data for Problem Identification
Revising Instruction
Content Revision
Procedural Revision
Summary

Student Objectives

The student who understands the material in this chapter should, with the book closed, be able to:

1. Design, label, and explain tables summarizing learning and observational data for use in revising an instructional activity
2. Identify the preinstructional and instructional components of an activity which may need revision, and the basis upon which revision decisions may be made
3. Describe an appropriate type of content revision given various types of performance data
4. Identify the various procedural components associated with the implementation of instruction which may need revision and the basis upon which revision decisions may be made
5. Describe appropriate types of procedural revisions which might be made given various types of student comments and teacher observations

**Summarizing Data
for Problem Identification**

We have now reached the last step in the systems approach model, which, as Figure 11-1 shows, is labeled formative evaluation. Actually, the formative evaluation process was first described in Chapter Seven in terms of developing various evaluation instruments which could be used in conjunction with an instructional activity. After an instructional activity has been completed, it is necessary to determine what went right and what went wrong and to try to revise the instructional process in order to make it more effective.

The revision of instructional materials and procedures is actually a two-step process. The first step is determining that a problem exists, and the second is correcting it. The process may be compared to driving an automobile: when the average driver has a flat tire, he is soon aware of it because it is so obvious. It is equally obvious to most drivers what they must do in order to rectify the problem. Similarly, in your instructional activities there may be obvious problems with equally obvious solutions. However, there may also be subtler types of problems, which are more difficult to detect and still more difficult to remedy. The driver of the defective automobile may hear a faint, peculiar sound coming from his engine and not be sure how serious the problem is—if there really is a problem.

FIG. 11-1 *Systems approach model*

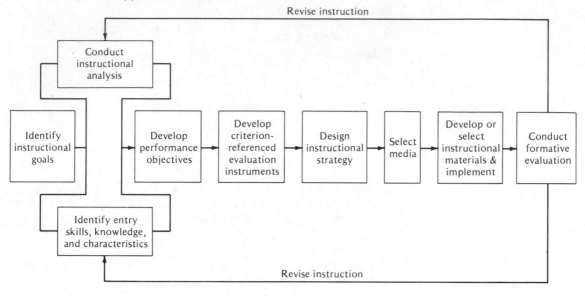

He may also be unable to decide how to solve it. He can either ignore it or take it to an expert mechanic. You, as the teacher, must be the expert mechanic who can look at these less obvious instructional problems and determine what should be done to remedy them.

We have previously noted that through the use of the systems approach to instruction, the teacher becomes, in effect, a behavioral scientist. Using our analogy to the automobile once more, we can say that a mechanic uses his knowledge and his tools to repair a malfunctioning automobile. As a behavioral scientist your tools are your experience with instructional processes and the data you collect on their effectiveness. This data can be an invaluable aid in identifying faulty instructional activities and suggesting alternative instructional strategies.

The first step, then, in the formative evaluation and revision process is to summarize the data available to you. In order to evaluate an instructional activity, you must somehow summarize the data for each part of that activity. Figure 11-2 shows a suggested format for summarizing your data. Look closely at the figure, and then we will discuss each heading so that you will know what might be included.

Along the left-hand margin of Figure 11-2 are the headings Objective I, Objective II, Objective III, and Entry Behaviors. Under each of these are numbers. Let's take Objective I. The table states that two evaluations were taken for Objective I. These could be

"... it is necessary to determine what went right and what went wrong and try to revise the instructional process in order to make it more effective."

two test items, two different assessments of psychomotor behavior, or two indicators of some social or affective behavior. The important point is that they are two assessments directly related to the student's achievement of Objective I. The numbers shown under Objectives II and III also refer to assessments of these objectives. As you will note, there is only one type of assessment under Objective III. Likewise, in the same column we have listed Entry Behaviors and numbered them 1, 2, 3, and 4. These represent the various assessments of the students' abilities to perform those

FIG. 11-2 *Data summary sheet*

	Pretest data		Learning data				Posttest data	
	$\overline{X}$	% criterion	$\overline{X}$	Short	Long	Performance	$\overline{X}$	% criterion
Objective I								
1	.00	.00	27	18	55	.85	.80	.80
2	.10	.10					.80	.80
Objective II								
1	.30	.30				1.00	.50	.50
2	.20	.20				1.00	.60	.60
3	.30	.30				1.00	.70	.70
Objective III								
1	.43	.43				.87	.50	.50
Entry behaviors								
1	.95	.92						
2	.99	.99						
3	.85	.90						
4	.93	.95						

skills which have been indicated as being necessary for entry into the instructional activity.

Now let's look at the column headings. The first two columns are labeled "Pretests." Under Pretest, the first column represents students' mean performance on the evaluation instrument ($\overline{X}$). The second column represents the percentage of students who have met the predetermined criterion level you have established. Therefore, if we look at the pretest information for Objective I, we notice that on the first evaluation question the mean performance was 0 percent and that therefore, none of the students met the pre-established criterion. Likewise, on the second evaluation item the mean performance was 10 per cent, and thus 10 per cent of the students met the criterion. The data for Objective II are similar. Students' mean performance on these pretest items was 30 per cent, 20 per cent, and 30 per cent. Likewise, 30 per cent, 20 per cent, and 30 per cent of the students met the preestablished criterion for that objective on the pretest. The rest of the data in this column can be similarly interpreted.

Students' posttest performance is indicated in the third pair of columns. The column headings are identical to those used for the pretest, namely, the mean performance of the total student group and the percentage of students who met the preestablished criterion level. Note that there are no posttest data on entry behaviors. It is unlikely that you would retest the students on their understanding or performance of skills required for entry into the instructional program after they have completed that program.

The second set of column headings shows the data which may have been collected *during* the instructional activities. For example, the first three columns pertain to the time required to study for a particular objective. The data show that the class spent an average of 27 minutes studying for Objective I, with individual students' times ranging from 18 to 55 minutes.

No learning-time data have been included for Objective II, but there are group performance data for each of the three evaluations that were administered during instruction related to that objective. The table indicates that 1.00 or 100 per cent of the students were able to perform successfully on all three evaluations of the objective. The class average for Objective III was 87 per cent.

It is important that you understand exactly what the numbers represent so that you can summarize your own learning data. Look now at Figure 11-3, which contains hypothetical data for the posttest performances of ten students. You can relate these data directly to the column on posttest performance in Figure 11-2. Figure 11-3 shows the performance of ten different students on the six different evaluation items for the three objectives. If we read down the right-

FIG. 11-3 *Hypothetical posttest performance for 10 students*

Students	Objective I		Objective II			Objective III	
	1	2	1	2	3	1	
S_1	1	1	1	1	1	1	1.00
S_2	0	0	0	0	0	0	.00
S_3	1	1	0	1	1	0	.67
S_4	1	0	1	0	0	0	.33
S_5	1	1	1	0	0	0	.50
S_6	1	1	0	1	1	1	.87
S_7	1	1	0	1	1	0	.67
S_8	0	1	1	0	1	1	.67
S_9	1	1	1	1	1	1	1.00
S_{10}	1	1	0	1	1	1	.87
	.80	.80	.50	.60	.70	.50	

hand column of the table, we can see the percentage performance of each of the students. You can use these data to assign a grade if necessary. However, at this point we are more interested in observing the data as they are summarized by objective rather than by student. Therefore, reading down the columns, it can be seen that 80 per cent of the students performed successfully on both the evaluation items for Objective I. That is, 8 out of 10 or 80 per cent of the students were successful on each item. Similarly, 50 per cent, 60 per cent, and 70 per cent of the students were successful on the three items for Objective II.

Since only 1's and 0's appear in Figure 11-3, we must assume that the evaluation items were scored as either right or wrong; i.e., they were either achieved or not achieved. This situation would most often be observed in a conventional paper-and-pencil test, in which each item is scored either right or wrong. However, the same type of summary table could be utilized if the criterion measurement were in terms of a percentage, number of feet, time, etc. An average performance for the group on each evaluation item could be determined, as well as the percentage of students meeting the criterion. Note that when 1 and 0 scoring methods are used, as shown in Figure 11-3, then the mean performance of a group of students on each objective item and the percentage meeting the criterion are identical. You should be cautioned that if several different types of assessments are used for the various objectives, then you cannot summarize a *student's* performance by simply adding up his various scores. That is, you cannot add test item performances to the number of feet he may have thrown an object.

The second type of data you can use in the revision process are the subjective comments and attitudes of students. This information may be in the form of a questionnaire regarding the procedures used in the course, or in the form of various comments made to you in formal debriefing sessions or in informal moments during the instructional process. Figure 11-4 shows one possible format for summarizing this type of information.

The type of activities in which you might be most interested, such as the pretest, the orientation to the instructional activity, the actual instruction, and the posttest, have been listed in the left-hand column. The next column simply represents the tally of the favorable and unfavorable comments about this aspect of the instruction. The final column is a listing of some of the statements which might be made in an interview or on an open-end questionnaire, with tally marks to indicate how many people made approximately the same statement. For example, in Figure 11-4, the tally marks indicate that four people said in essence that they were glad to take a pretest in order to find out how much they knew before the course actually started. One person indicated that he thought the test items were unfair. Since there are many ways of collecting this kind of subjective reaction from students, there are also various ways of summarizing it; Figure 11-4 shows one method of doing so. In general, though, you should try to cluster or group various students comments in order to relate them to the content and procedural aspects of an instructional activity.

Revising Instruction

Content Revision

The procedures for revising instruction will be divided into two sections. The first will be called content revision, and the second, procedural revision. As we describe these two types of revisions, you will find that the distinctions between them are sometimes blurred. The distinction is not critical, but it helps to emphasize the fact that student performance may be influenced by two factors. The first is the obvious one, namely, the quantity and quality of the instruction and the planning that has gone into it. The second is a more mechanical, but perhaps equally important, one, which includes the actual procedures used in implementating an instructional activity.

Earlier in this chapter we used the analogy of the automobile mechanic and the teacher who attempts to revise his instructional materials. Another analogy may be meaningful at this point. The proverbial blind man would have to touch an elephant in many different places before he could begin to get some idea of what the elephant was really like. The teacher who attempts to revise instructional activities is in somewhat the same situation. He must collect a great deal of data before he can determine exactly what the problems "look like." The procedures described below, which are based

FIG. 11-4 *Sample format for summarizing subjective information*

Activity	Frequency Favorable	Unfavorable	Comments
Pretest	8	2	//// Good to know where I stood before course started.
			/ I thought items were unfair.
Orientation	5	5	
Instruction	7	3	
Posttest	10	0	

on experience with a variety of different procedures, are a systematic set of steps you may follow in order to identify and remedy faulty areas of instruction.

Test Items   One of the most important pieces of information in the revision process is the evaluation data from the various instruments which have been employed in the instructional process. As a first step in the revision process, you should reassure yourself of the validity of the test items you have administered and note any inconsistencies between them and the objectives which were originally established for the instruction. You can now obtain a rough indication of the reliability of various items by examining the relationship among those which have been used to measure the same objective. Your instrument will tend to be reliable if those students who got one item right for an objective tended to get other items right also. Those students who tended to miss one item should also tend to miss other items related to the same objective.

This first step is very important because all the remaining analysis will be based upon the data which are derived from students on these test items. Therefore, if the items themselves are invalid or unreliable, then any decisions made on the basis of data obtained from them will be also.

Entry Skills   The second step in the revision process is to examine the data obtained from the students concerning their entry skills. Hypothetical data on such behaviors are listed in the bottom portion of the table in Figure 11-2.

Obviously, what you want to determine at this point is whether the students actually have the skills you assumed they would have when they started the instruction. The data in Table 11-1 would suggest that for this hypothetical task the large majority of the students did, in fact, have the required entry skills. Therefore, you can be assured that any problems the students may have had in achieving the terminal objectives were due to some form of faulty instruction. However, if a significant percentage of students did not have the required entry skills they would also have considerable difficulty with the instructional program. This would be a clear indication that with subsequent groups you should provide instruction for those students who do not have the required entry skills. The other obvious alternative would be to exempt students who lack the required entry behaviors from this activity and place them in a more appropriate one.

An additional observation might be made with regard to entry skills. If the average performance of students who had most or all of the required entry behaviors differs greatly from the performance of those students who did not, then the identified entry behaviors are probably critical to the successful performance of the instructional task. However, if the mean final performance of the two groups does not differ to any great extent, then it suggests that not all the entry skills are necessarily prerequisite to success in the learning activity. Certainly, the entry skills test should never be used as a criterion for admitting students into the instructional activity until the test has been fully validated for this purpose.

Pretest Data   After you have determined the validity and reliability of evaluation items and examined the data from the assessment of entry skills, your next step is to look at the pretest and posttest data. If students' performance on the pretest is low, i.e., if the average score is 20 per cent or less (or less than 10 per cent of the students achieve the criterion for an objective), then it would appear they did in fact receive instruction in something they could not already do. However, if their performance on the pretest is quite good, this suggests that they already had the skills for which they received instruction. This clearly indicates that you should administer the pretest prior to an instructional activity so that you can provide students who already have the skill with some alternative instructional activity or use them as alternative instructional resources with other, less-skilled students. Furthermore, high pretest scores make it almost impossible to evaluate instruction; the students already have the skill so there is little that any instruction could add to it.

With many of the activities you are likely to teach, it is more than likely that you will find a scattered pattern of results. In general, the pretest will show low student performance, but here and there you may find various types of evaluation data indicating that the students have already acquired some of the skills you may wish to teach. These particular skills should be taken into account along with the information which is derived from the posttest data.

Posttest Data Individual item scores and total scores from the posttest are certainly the best indicators of success or failure in achieving instructional goals. Often a teacher will set a goal of 90/90 for his instructional activity, i.e., his goal is that at least 90 per cent of the students will achieve at least 90 per cent of the objectives. He can summarize the posttest data to determine whether his goal was realized. However, for revision purposes it is more important to look at both the pretest and posttest scores for each of the subordinate objectives as well as for the terminal objective.

One of the most effective ways of examining pretest and posttest data for the various objectives is in the context of the learning hierarchy which was developed for the instructional activity. Figure 11-5 shows the structure for a hypothetical terminal skill which has six subskills. Let's assume that the data for one or more objectives for each of these skills have been collapsed into a pretest and posttest score for each skill. We can now analyze the instructional process in terms of the students' success or failure as they experience it. Let's assume that based on the hierarchy in Figure 11-5, the teacher would teach these skills in the sequence A, B, E, then C, D, F, and then the final terminal skill, G.

The pretest scores for skills A and B indicate that relatively few students could perform the skills prior to instruction but that essen-

FIG. 11-5 *Pretest and posttest data superimposed on learning hierarchy*

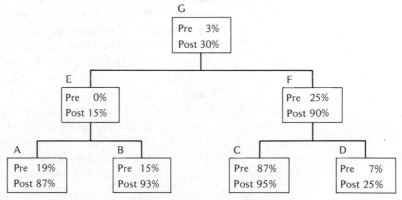

tially all the students could perform them afterward. Therefore, it appears that the instruction was beneficial. However, the data for skill E indicate that while none of the students could perform the skill on the pretest, neither could very many of them perform it on the posttest (15 per cent); so that apparently the instructional activity failed to teach that skill.

Now, looking at skills C and D, we find that almost all the students could perform skill C on the pretest and posttest; however, for skill D, there was very little student growth from the pretest to the posttest. It is interesting to note that in our hypothetical example, 90 per cent of the students were able to achieve skill F, even though only 25 per cent could achieve skill D. These results suggest that there was a great deal of transfer from skill C to skill F and that the inability to perform skill D did not interfere to any great extent with the learning of skill F. There is some question about the relevance of skill D to the performance of at least skill F, if not to the performance of the terminal skill.

Now we can look at the relationship between subskills E and F and the terminal skill, G. It would appear from the data that this example is consistent with the positive transfer theory which underlies the instructional analysis process. While 90 per cent of the students were able to perform skill F on the posttest, only 15 per cent could perform subskill E. Subsequently, only 30 per cent of the students were able to achieve the terminal skill, G. The data suggest that perhaps 15 per cent of the students who could perform skills E and F positively transferred their learning of those skills to the learning of terminal skill G. Perhaps another 15 per cent of the students who were not able to perform skill E received enough instruction on skill G so that they could in fact overcome their lack of skill E. However, it should be noted that only a small percentage of students were capable of doing this.

On the basis of the data presented in Figure 11-5, what tentative conclusions can we now make about the instructional activity upon which the data are based? The most obvious, perhaps, is the fact that only 30 per cent reached the desired criterion level of the terminal objective. Therefore, it would seem that there were significant problems with the instruction the students received. If we consider the first level of subordinate skills, it would appear that the instructional activities associated with skill E are certainly a prime target for further consideration, because only 15 per cent of the students were successful with that skill. We can be even more certain of that diagnosis when we look at the skills immediately subordinate to skill E, namely, A and B. Here the posttest performance was very high and therefore should have provided a positive transfer to skill E.

If we examine skills F, C, and D, we can see that there is a good, solid relationship between C and F. The students' fine performance on each suggests positive transfer. However, how do we explain their relatively poor performance on skill D if they were able to achieve skill F? If we should look closely at skill D, and particularly at the objectives and test items associated with it, we might find, for example, that it was not a logical subordinate skill that would be associated with skill F. Or, to the contrary, we might find on reanalysis that skill D was a higher-order skill than F and that they should actually be interchanged in the hierarchy. If you look at Figure 11-5 and mentally interchange the data for skills D and F, you will find that the data are more nearly consistent with the positive transfer theory of the learning hierarchy. That is, if students performed well on skills C and F but seemingly could not then achieve skill D, perhaps skill D was poorly taught. The faulty instruction associated with skills D and E explains why there was such poor posttest performance on the general terminal skill.

The foregoing analysis suggests that the next step in the revision process is reevaluating the objectives which have been derived from the learning hierarchy. But first, let's see if you understood our discussion of using pretest and posttest scores in the learning hierarchy to identify potential areas for revision. Figure 11-6 includes a set of summary data for four skills along with a hypothetical learning hierarchy for those skills. Place the summary data in the appropriate areas in the hierarchy, and then, in the space provided on page 270, state what happened in this instructional activity. Did a significant proportion of the students reach the terminal objective? If not, why not?

FIG. 11-6 *Performance data summary and learning hierarchy*

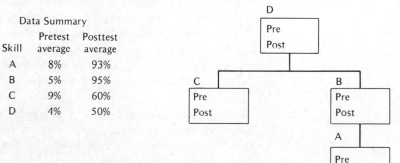

Data Summary

Skill	Pretest average	Posttest average
A	8%	93%
B	5%	95%
C	9%	60%
D	4%	50%

Performance Objectives   After examining the pretest and posttest data in the learning hierarchy, you will want to reexamine the performance and subobjectives which you used in designing your instructional activity. Reexamining the objectives is analogous to reviewing test items. It is, in essence, a validity check; i.e., do the objectives continue to represent the kinds of performances which will be expected from the students?

In reviewing the objectives, you should refer to data like those which were presented in Figure 11-2. For example, you might find that students performed extremely poorly on an objective because it was not stated clearly or did not accurately represent the subordinate skill in the learning hierarchy. You should review all the objectives systematically to insure their continuing validity and, if necessary, revise them in light of the performance data you have obtained and your analysis of the learning hierarchy and entry behaviors.

The four steps in the revision process we have suggested thus far are shown graphically in Figure 11-7. The model indicates that after you have reviewed the evaluation items, your next step is to examine the entry behavior scores and the pretest and posttest performances within the context of the learning hierarchy. You will use this information when you reexamine the performance objectives. You are now ready to revise the instructional content and the instructional procedures.

Instructional Activity   At this point we are ready to undertake one of the major tasks in the revision process, namely, evaluating the instructional activity itself. By instructional activity we refer specifically to those things which you might do in order to convey the content of the instructional goal, such as lecturing on, discuss-

FIG. 11-7 *The revision of systematically designed instruction*

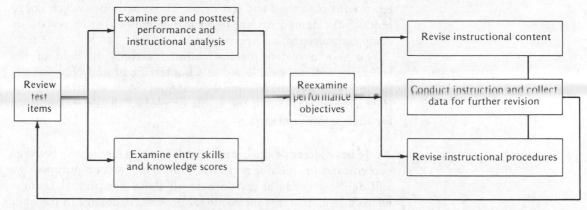

ing, or modeling a particular kind of behavior; *and* the instructional media you might use, such as textbooks, video tapes, programmed instruction texts, etc. The term instructional activity should be understood to mean the instructional message, regardless of its source.

In order to evaluate objectively the effectiveness of an instructional activity, it is necessary to again review (a) pretest and posttest learning data, (b) any other learning data which are available, (c) students' comments, and (d) your own observations. As a start, examine the posttest performance on each objective. If it was relatively poor, but there are some indications that the students did in fact master the objective during the learning activity, then there is probably a need for additional instruction or perhaps review exercises to insure the retention of the information or skill.

If the data indicate that performance on a particular objective was quite poor on the pretest, during learning, and on the posttest, then you should completely reassess the instructional activity. These types of data immediately suggest that the instruction is not consistent with your stated objective and the method you have chosen for evaluating it. In their comments about the course students quite often reveal the existence of this type of problem. When they are informed in advance of the objectives of a unit of instruction, they know when the instruction is not relevant to those objectives.

A third condition you should not overlook is that of significantly good performance on the pretest, during learning, and on the posttest. This suggests that the students do not need the instruction and that it could perhaps be reduced in scope or totally eliminated.

In general, you should reexamine each instructional activity, regardless of the medium though which it is presented, for consistency with objectives and test items. Those activities which fail to produce the desired student performance may either be revised or completely deleted, and new activities added.

You should review the instructional activity in light of the general guidelines established in Chapters Eight and Nine for designing effective instructional strategies and selecting media. Instructional events and the media employed should be reviewed for their appropriateness.

Procedural Revisions

As we have already noted, it is often difficult to distinguish between content and procedural revisions. However, for our purposes we will define procedural revisions as all those operational changes which are made in the instructional process. A change in the hour of a class meeting would be a procedural change, as would a shift from 16 mm films to a tape/slide presentation.

There are few data to indicate the effect of procedural changes on the learning process. However, it is quite clear that many changes should be made in order to improve student performance. Such changes may save learning time or improve students' attitudes toward the subject matter or the mode of instruction. There is little evidence to support the belief that because students like a particular type of instruction, they will learn better with it. In fields such as physical education the attitudes students develop toward instructional activities are as important as the skills they learn and retain.

The two major sources of data which are usually employed to determine the need for procedural revisions are teacher observations and student comments. While these types of data may not be considered "hard" or terribly objective, they often provide ideas for relatively simple changes which can greatly modify the effectiveness of instructional activities. This type of information is advocated by many designers of instruction, who have developed procedures which appear to be very good in theory but which are almost totally unworkable in the learning situation.

One way of systematically considering the various types of procedural revisions is by following the format presented in Figure 11-4, which includes comments about pretesting, orientation, instructional activities, and posttesting and grading. In the following paragraphs we will consider each of these in turn.

Pretesting You can use pretesting data in evaluating students before they enter the instructional activity; in placing them at appropriate levels in that activity; and in revising instruction. Because of the importance of the pretest data, you should review the pro-

". . . you should reexamine each instructional activity, regardless of the medium . . . for consistency with objectives and test items."

cedures used to collect them and their use in the instructional activity. Since this type of testing is not commonly employed in instructional programs, students may have difficulty understanding its meaning; therefore, you should be sure to emphasize the importance of their full participation in the testing program.

Orientation to the Instructional Activity   Many of the instructional activities implemented by teachers using the systems approach are quite unique and novel in comparison with those which have been used in the past. Therefore, one of the single most important requirements of successful instruction is that students be properly oriented to the instructional activity. The procedures which will be used should be completely explained, particularly any individualized aspects of them. In addition to describing the general flow of the course, you should inform students of the type of instruction, evaluation, and grading they will receive. The criterion-referenced system will also be somewhat new to students and will require a great deal of explanation. If you consider the amount of time and effort you have invested in reading and studying this textbook in order to understand the new ideas it presents, you will be able to appreciate the position of the student, who has not had this type of instruction or a variety of experiences in the past.

Students may also be unaccustomed to being asked to provide the teacher with feedback about the instructional activities. Students should be made aware that you are continually trying to improve instruction and that their constructive comments will help to make an instructional activity more efficient and effective for the next group of students.

All these factors are important in the orientation process; and therefore, you should take them into consideration if you revise that process. An important part of any orientation can be an elaboration of the misunderstandings experienced by students who have previously participated in the instructional activity. Thus a log of such misunderstandings can make an important contribution to the revision process.

Instructional Procedures Students' comments can be of prime importance in evaluating instructional procedures. For example, their comments may indicate that the teacher's role in the instructional process should be modified. They may see a need for more discussion of particular topics, more modeling of particular kinds of activity behaviors, more feedback on individual performances, or additional information about concepts they don't understand. A common observation about self-paced instructional activities is that many students procrastinate in learning them. While this may be attributed to human nature, it may also indicate that students are redirecting their energies to requirements in other courses they are taking at the same time. If procrastination is in fact a problem, then it may be helpful to insert intermediate deadlines in the activity so that students will pace themselves more effectively.

Numerous suggestions of this type might be made; however, their relevance to any particular instructional activity would only be hypothetical. The point is that you should depend on your own observations of students' progress in the various instructional activities. Students can make valuable suggestions for improving the learning setting, either while instruction is taking place or afterward in a structured interview. An in-depth analysis of such proposed procedural changes might indicate that they conform quite well to what we know about learning theory. Or changes may be based on straightforward pragmatic decisions, which must be made in order to enhance learning.

Posttesting and Grading The two final areas of concern in the revision process are the administration of the posttest and the assignment of grades, procedures which warrant constant review because of their significance to both you and your students.

You will have to decide at what point a student should receive an evaluation of his performance on the terminal and subordinate objectives. Should each objective be evaluated only once, or should there be review evaluations and an end-of-course evaluation? Should a student be permitted to take his end-of-course evaluation whenever he completes the course, or only when all the other students have completed it? The latter question pertains only to individualized instructional activities. However, it is relevant to the revision process in terms of the delay which occurs between the end of the learning activity and the evaluation of it. You should certainly take this matter into consideration when reviewing the instructional procedures.

Various methods have been suggested in Chapter Seven for assigning some type of evaluation or grade to a student's performance when criterion-referenced evaluation is employed. The effectiveness of the procedure selected should be reviewed and discussed with the students. Do they understand and accept the procedures, or would they prefer some other method of grading?

One common method of assigning grades in a criterion-referenced evaluation setting is to preestablish point totals which students must achieve in order to obtain various grades in the course. Students should be informed of these totals before they engage in the instructional activity. If such a method is used, then the distribution of the actual number of points accrued by students should be reviewed following each instructional activity. You should determine whether the preestablished point totals are realistic in terms of the students' capabilities. The point totals may in fact be too high or too low. Only repeated experience with the learning activity and the spread of student abilities will enable you to make a fair assessment of these criteria.

Summary

This chapter has described the final step in the systems approach to the teaching of physical education. Formative evaluation and revision are the link between student performance and the overall design of the instructional activity. This process is not well understood or documented. Various methods were described for summarizing what may sometimes be viewed as vast amounts of data and comments in a format which will reveal any weaknesses in an instructional activity.

A distinction was made between revising the content of an instructional activity and revising instructional procedures. In revising content, the first step is to reexamine the evaluation items and assess student performance on the pretest, posttest, and entry behaviors. Next, the performance objectives should be reexamined, and finally the content of the instructional activity itself should be revised. The procedural revisions are related to pretesting, orientation, instruction, and posttesting and grading. In each case, an attempt was made to call attention to potential problems in each of these areas and to suggest alternative techniques for resolving them.

Resources

Books

Education

Espich, J. E., and Williams, B. *Developing Programmed Instructional Materials.* Palo Alto, Calif.: Fearon, 1967. Discusses use of student performance data to revise programmed instruction materials.

Gerlach, V. S., and Ely, D. P. *Teaching and Media: A Systematic Approach.* Englewood Cliffs, N.J.: Prentice-Hall, 1971. Chapter on evaluation includes a discussion of the use of feedback for instructional revision.

Markle, D. G. *The Development of the Bell System First Aid and Personal Safety Course.* Palo Alto, Calif.: American Institutes for Research, April, 1967. Describes an empirical revision process and the success achieved by successive groups of students.

Rayner, Gail T. *A Revision Methodology for a Computer-Managed Instruction Course: An Empirical Approach.* Ph.D. dissertation, Florida State University, 1972. Design and validation of one procedure for revising instructional materials.

Articles

Education

Brooks, L. O. "Note on Revising Instructional Program." *Psychology Reports,* 20:117–118, 1967.

Brooks, L. O. "Response Time During Instruction." *Perceptual and Motor Skills,* 25:203–204, 1967. Describes the use of learning time (response latency) as a factor in the revision process.

Dick, W. "A Methodology for the Formative Evaluation of Instructional Materials." *Journal of Educational Measurement,* 5:99–102, 1968. Describes the collection of revision data by an evaluator who was not participating in the design and development of the instruction.

12

Accountability and Physical Education

If you met a friend on campus and asked him where he was going, he might say he was just taking a walk and going nowhere in particular. Your reaction might be to join him or at least envy him his pleasant break from the routines of campus life. However, if you met him on three or four more occasions and he continued to say he was simply walking around, was going nowhere in particular, and had nothing to do, you might become concerned about his full participation in the academic program, his general educational goals, and whether he was achieving them.

Many people are beginning to feel that education has been "taking a walk" for too long and are raising serious questions about what is going on in education. There appears to be an increasing national demand to know where education has been, where it is going, and how effective it has been. In other words, there is an attempt to get educators to answer the question, "What are we getting for our buck?"

Most of the questions now being asked can be lumped under the term accountability, or more specifically educational accountability. In the past such questions about education were successfully answered, primarily by administrators, in terms of the number of buildings erected, the size and diversity of schools and staffs, and the number of graduates from various programs. The really "successful" public schools were able to boast that a large number of their graduates had gone on to Ivy League or Big Ten colleges. However, such answers are no longer acceptable. More and more, the public is expressing a need for full disclosure of what pupils are able to do when they enter schools, what they are able to do when they leave, and how much it costs to produce these changes in behavior.

In this chapter we will briefly discuss the accountability movement and why we feel that the skills described in this book are critical to the professional accountability of the modern physical educator.

277

Chapter Contents

What Is Accountability?
Who Is Accountable?
Why the Accountability Movement?
When Will Accountability Plans Become Operational?
Accountability and the Systems Approach
Teaching as Behavioral Technology
Change in the Schools
Summary

Student Objectives

The student who understands the material in this chapter should, with the book closed, be able to:

1. Identify at least three of the primary components of an accountability system
2. Identify the persons and/or agencies who should be accountable in an educational system
3. List the major problems associated with the establishment of an accountability system
4. Describe the relationship between educational accountability and the systems approach to teaching
5. Describe what is meant by the term behavioral technologist

What Is Accountability?

There is no textbook definition of educational accountability which is agreed upon by all the experts in the field. It is a very global concept which depends on a variety of techniques for its implementation. Many people believe that simply applying these techniques to the instructional and administrative processes will enable educators to account for the use of public funds.

Robert Garvue (1971) has defined the concept of educational accountability as follows:

> The concept of educational accountability is concerned basically with techniques to guarantee a certain level of student performance relative to stated objectives and goals with an accompanying efficient use of resources. The educational enterprise will face growing competition for resources, with ecological, health and continuing national defense demands among others being highlighted. A criterion of effective fiscal support for public education will be the degree of faith that the citizenry has in the ability of school systems to "give the people their money's worth." (p. 34)

In connection with this example of accountability and the competition for public funds, it is interesting to note that when a person enters a hospital with a broken leg and later leaves with his leg mended, it is obvious to the layman that the hospital has achieved a publicly acknowledged goal, although he may be quite concerned about the cost involved. On the other hand, far fewer "cures" take place in prisons and mental hospitals. Administrators of such institutions are now indicating that they have insufficient resources to do their job adequately. This complaint is well known to educators, who have used it effectively in the past. The difference seems to be that people are beginning to acknowledge their lack of support of penal and mental institutions but feel they have in many respects supported education. They are now asking that their support be accounted for.

Who Is Accountable?   You should not be surprised to learn that the supporters of educational accountability are not generally referring to their own performance, but someone else's. When discussing the topic, they usually refer to "educational" accountability or the accountability of "the schools." An experienced teacher knows that the new responsibility will be transmitted from the superintendent to the principal, and ultimately to the teacher. Thus the demand for teacher accountability will certainly spur ambivalent feelings. On the one hand, educators have sought for years for ways to evaluate teachers' skills and effectiveness in order to reward those who are most successful (and perhaps, conversely, to penalize the unsuccessful). However, circumstances vary greatly from one classroom to the next, and it is hard to believe that criteria could be developed which would apply to all teachers.

If the concept of accountability is to be viable, it must certainly include all the components of the educational system and those agencies which support it. Ruth Chapman (1972) has effectively included many diverse groups in her description of educational accountability, which will be utilized in the remainder of this chapter:

Accountability, which is getting a large dose of exposure these days in education journals and editorial columns, shouldn't be restricted to one-directional focus.

It's a two-way street. In fact, it's a four-way intersection and it would be well to note the stop signs at each road, and, at least, to pause.

We're very concerned, and properly so, that we know what we're getting for the money we're investing in our educational system. No argument there.

Schools should be accountable, we say. Teachers should be accountable. Students must learn and we should have some measure of how much they learn and how much it costs.

That's oversimplifying it, but the idea is there.

> The only trouble is that all we ever seem to hear or talk about is the accountability that's attached to school systems, and to teachers.
>
> But there's a lot more accountability around that needs to be admitted and accepted and directed.
>
> Government is accountable—to children, as well as to voters.
>
> Superintendents are accountable—to children as well as to taxpayers.
>
> Legislators are accountable—to children, as well as to taxpayers.
>
> Parents are accountable—to children, as well as to neighbors and status and ambitions.
>
> Scholars are accountable—to children, as well as to attainment of high levels of knowledge.
>
> Society is accountable—to children. (p. 4-D)

Dissatisfaction with present programs, curricula, and classes is evident at all levels and in all areas of education, and suggestions for strengthening the instructional process have been taken from various sources. In the May 1972 issue of the *Journal of Health, Physical Education, and Recreation* are a number of articles with recommendations for upgrading instruction. Some of the letters to the editor contain ideas about a competency-based skills program for physical education majors. This program, being developed at the University of Texas, El Paso, will include specific teaching behaviors for enabling students to achieve required competencies. The competency-based approach (versus the traditional time- and course-centered approach) is viable at all levels of education and in a variety of programs.

In the same issue of *JOHPER* Stratton Caldwell calls for more humanistic physical education. He advocates self-actualization through movement experiences, which has direct implications for the training of teachers and, in turn, the development of their students. He criticizes both teachers and teaching methods:

> To achieve this goal means the elimination of the jock, the throw-out-the-baller, the military martinet, the professional who died on the vine years ago, the incompetents and uncaring who masquerade as teachers while fostering student passivity, obedience, conformity, and dependence. This will occur only when students, administrators, tax payers, and *we* as caring, concerned teachers simply refuse to accept programs, practices, and procedures that poison us all. (p. 32)

These are strong words, but they come from a concerned physical educator who is deeply committed to high-quality physical education experiences. In another article, Thomas Schechan and William Alsop attempt to show how sport can be used as an effective medium for positive change in cognitive, affective, and psychomotor behaviors. Their analysis of sport as an educational process is consistent in many ways with the material presented in this text. In other articles scholars express concern about graduate education, professional preparation curriculums, and student teaching experiences.

As this single issue of *JOHPER* reveals, physical educators are tending increasingly to (a) boldly identify weaknesses in programs, and (b) recommend desirable alternative strategies for remedying them. The commitment to educational goals, individual students' achievement levels, learning processes, and the like reflect the physical educator's realization that he must assume a greater responsibility, or accountability, for his actions. There is no one way of becoming accountable, but concern, interest, sensitivity, effort, and knowledge help.

Why the Accountability Movement?

Leon Lessinger, the man most identified with the education accountability movement, has identified some of our most obvious failures. For example, approximately one out of every four American children drops out of school after the fifth grade and before graduating from high school. In addition, one out of every four 18-year-old males fails the mental ability tests for entry into the armed services. Lessinger (1971) has described the seventies as an era of educational trauma, in which public education has been and will be called upon to defend itself against attacks from many quarters.

Sadly, impatience with these failures can have unfortunate consequences. More and more frequently, programs are being cut back or even totally eliminated. The cost of education is extremely high, and support is increasingly difficult to obtain. It is little wonder, then, that educators have been required to demonstrate the contributions of all aspects of education to worthy educational goals; in fact, to justify the very existence of their programs. It has always been difficult for physical education and sport to gain acceptance, and that continues to be the case. Our programs and objectives are severely challenged in certain circles, physical education requirements are increasingly eliminated, and programs are being deemphasized.

Accountability in physical education must now go beyond superficialities—the generalities of program outcomes. Teachers must know more about their specialized content, the teaching-learning process, and ways of demonstrating positive changes in students. Many citizens are attacking various facets of education, and physical education and athletic programs have proven to be more vulnerable than many others. These programs are costly, and the expenses incurred—for equipment, facilities, the salaries of teachers, coaches, and maintenance and supervisory personnel—must be justified.

". . . public education has been and will be called upon to defend itself against attacks from many quarters."

Parents want to know—to be shown—that their children are fulfilling their potentialities, that the time and expenses involved in physical education and athletic programs are indeed worthwhile. Since the parents pay for education, they have every right to question educational practices and curricula. It will not help physical educators to evade responsibility, to attempt to escape, to conceal weaknesses in teaching, to defraud the public, or to place the blame elsewhere. A more positive approach would be to improve methods to satisfy educational requirements, to help students achieve objectives, and to legitimately justify instructional procedures.

When Will Accountability Plans Become Operational?

Educational establishments have not been completely oblivious to the developments Lessinger has described. Action has been and is being taken. State departments of education and individual public schools are beginning to develop their own objectives for students and are drawing up plans for evaluating the achievement of those objectives. However, most realistic assessments of the present situation would indicate that we are a long way from developing a satisfactory system of accountability. There is the almost insoluble problem of training teachers to develop and/or collect objectives, to provide meaningful learning experiences related to them, to assess student performance with evaluation instruments, and to develop appropriate analysis and reporting procedures. Preservice and inservice training of teachers would have to be undertaken on a national level. William Deterline (1971) has written:

> The teachers, the instructors, in an accountability setting, have to function effectively as tutors, diagnosticians, remediators, managers, counselors, advisors, conversationalists and stimulating consultants. These skills are not part of most teacher-training curricula. Some will have to be learned by experience on the job, and some people will be able to develop higher levels of proficiency at one skill than another. Eventually, we will have to learn how to analyze and teach these skills, and be accountable for doing so, to at least minimally adequate level. (p. 19)

Although educators in general may be slow to respond to the need for accountability and the establishment of standard criteria, there is a greater urgency to do so in certain educational areas. When states, communities, or school boards reanalyze funding priorities in a period when money is tight, as it is now, philosophies and emphases change. The extent and quality of programs like physical education are threatened. Although examples may be found in many parts of the United States, that of the State of Illinois reminds us of the urgent need for members of our profession and all individuals connected with us to work intensively to provide the accountability necessary to establish untarnished respectability.

During the 1971–72 school year, the state-supported colleges and universities in Illinois were threatened by budget limitations and program priorities established by the State Board of Higher Education. This unfortunate circumstance created a tremendous stir statewide and nationally. Professionals pooled their talents in a united front to counter proposed changes, realizing that initial cutbacks in physical education programs could lead to further cutbacks. What happened in the State of Illinois could happen elsewhere.

Why wait until such a situation arises before taking action? Would it not be easier to take the initiative and minimize the probability of these events? Higher-quality programs will help to satisfy students as well as the public. Although it is beyond the scope of this book to recommend a standard procedure for the entire physical education profession or even for one teacher, we feel that the systems approach presented in this book could help teachers meet certain criteria of the accountability process.

Accountability and the Systems Approach

In our discussion of accountability you have undoubtedly recognized many of the concepts and terms which have been used elsewhere in the text with reference to the systems approach to teach-

". . . state-supported colleges and universities in Illinois were threatened by budget limitations and program priorities. . . ."

ing. We hope that you have concluded, as we have, that the skills required of the teacher in applying the systems approach are the same skills he will need to account to a supervisor for his use of resources—i.e., the ability to state instructional objectives and collect data on student performance. Without these skills and the ability to provide the appropriate learning experiences, the teacher is like the student encountered on campus who cannot describe where he has been or where he is going.

The skills described in this text do not make the teacher infallible, but they provide him with a starting point. Techniques must also be formulated by which administrators can assess the cost of individual learning outcomes and report these data in a meaningful way. These procedures must be developed with the full knowledge and cooperation of teachers and administrators in order to meet the needs of society. Deterline (1971) comments:

> . . . Accountability must start at the top, and must be applied to everything that has an instructional or instructional management function. The purpose is not punitive; the purpose is—to use a manufacturing term—quality control. The systems approach . . . is basically a methodology which requires exact specifications of what is to be accomplished by each component, and the use of techniques to see that each specification is achieved; meeting specifications is quality control. Seeing to it that specifications and quality control work—at all levels—is accountability. (p. 17)

Teaching as Behavioral Technology

The design of this book enables the teacher to become a behavioral technologist. In this era of technological revolution, do not think of the educational process only in terms of machinery and hardware. There is another technology to consider, behavioral technology, which refers to the learning-teaching process in which instruction is based on explicitly stated objectives, media for reaching those objectives, and assessment techniques for determining the effectiveness of the system. The science of shaping or modifying behavior is intricately woven into the art of teaching.

In this book we have attempted to formalize the teaching-learning process in a structured system. Accountability, at least as far as the physical education teacher and student are concerned, is facilitated when an instructional blueprint is developed. The systems model provides the guidelines. The teacher as engineer, or behavioral technologist, can be viewed as the designer of instruction who considers situational constraints and opportunities related to objectives and plans the most effective means for fulfilling those objectives.

We live in an age of automation. Unfortunately, automation implies using machines to replace people, and consequently has unappealing connotations for many educators and students. Although behavioral technology suggests the use of instructional media (e.g., machines), it is really the analysis of a particular situation that results in the best of all instructional approaches, with or without media. The emphasis on automation and man-machine interactions in industry has stimulated two new ways of thinking about the instructional process.

First of all, program and class objectives are categorized and analyzed and learning media developed within specified models. The model developed in this book is one such example. Thus a streamlined, efficient way of viewing the instructional process has been adapted from industry.

Second, the possibilities of using hardware and software in the instructional process have been seriously deliberated. Where automation is effective, it should be used. Rather than depersonalizing education, the use of machinery can have many advantages for the student and the teacher and can make more individualized teaching possible. Clerical functions can be reduced. Students can learn independently and at individual rates. Teachers can spend more time on individual students' problems and, in fact, provide more of a personal touch.

The teacher as behavioral technologist is therefore not another cold, calculating machine, designed to dehumanize education, but rather a learned, sensitive, and organized manager of instruction. The systems approach allows him to view more clearly the fruits of his labors, to evaluate them, and to establish accountability and credibility.

Change in the Schools

To change conventional practices is no easy matter. The adoption of the systems approach and other innovations in the schools challenges the vast majority of teachers and administrators, who have experienced and perpetuated more conventional educational methods. Resistance to change should be expected. Most of us would rather maintain familiar, comfortable routines than attempt what might appear on the surface to be revolutionary.

But change can and does occur. It can be caused by forces within or outside of the educational structure. It can affect the entire program or a small portion of it. It may be required in apparently harmonious situations where morale is high, or in extremely depressing circumstances. In any case, change should be initiated when a need is perceived, when curriculum outcomes can be improved upon. The process is easier when the personnel involved have modified their attitudes and are prepared to try new instructional approaches. Coercion and force do not usually result in productive programs.

Careful, rational planning by administrators and faculty, all contributing to desired goals, facilitates the process of change. If the systems approach is to work effectively, emotional upheavals among teachers should be kept at a minimum. The advantages of this approach over others should be objectively brought to the attention of the teachers involved. Discussions would help. Mutual planning would be beneficial, although leadership is important, as it is in any group situation.

It is possible for one teacher to use the systems approach in a structure in which other teachers are not doing so. Ideally, however, the systems approach should be used by an entire department or school. Students would adjust more easily, and the entire school could probably be run in a more coordinated fashion. In terms of implementation, however, it is obviously easier to change a small portion of an established program than to modify it in its entirety.

Teachers, like other people, resist change for various reasons. They may distrust or fear the motives of the person or persons proposing curricular changes and be uncomfortable with the perceived effects on their teaching roles. Insecurity often leads to an inability to try new things. Comfort and habit speak against change, for why change if one is satisfied? There are many reasons for not changing, as well as a number of ways of inducing change.

This book is all about change. It suggests a change in the instructional format and offers a bold alternative to current practices. Changing to a systems model should encourage the use of specifically stated objectives and should promote greater dependence on evaluation and accountability information, better understanding of learning research and theory and the teaching-learning

". . . the systems approach and other innovations in the schools challenges the vast majority of teachers and administrators. . . ."

process, more careful planning, more opportunities for individualized instruction, the use of multimedia, and the like. The systems approach can and will work. Its usage, acceptance, and success will depend greatly on your and your colleagues' efforts in planning for the new physical education.

Summary

Each employee who is paid from public funds is accountable to someone. Therefore, it is almost impossible to disagree with the general concept of educational accountability. We have tried to indicate that all teachers will have to be prepared to cope with it in the future. These coping behaviors may take the form of passive resistance, open hostility, or constructive participation in the development and implementation of new ideas. We have stressed that accountability is a two-way street. Teachers cannot be held accountable when they lack the resources to provide necessary learning experiences. Therefore, if predetermined objectives are to be met, the public should be informed of the successes of the educational program and the needs which still exist.

In the following quotation Floyd Christian, Florida Commissioner of Education, indicates five ways in which the roles of personnel and institutions must change in order to be accountable. In each of the commissioner's statements which starts with the word "it," if you substitute the words "systems approach" for "it," you will find that educational accountability and the systems approach are in many ways synonymous.

Accountability means different things to different people. I like to think of accountability as a concept that comes to grips with the notion that schools and colleges shoulder the responsibility for the learning successes or failures of their pupils. This concept calls for a revamping of much of our thinking about the roles of educational personnel and educational institutions at all levels:

It links teacher performance with student performance.

It implies precise educational objectives.

It forecasts the measurement of achievement.

It means, in effect, the schools and colleges will be judged by how they perform, not by what they promise; by their output, not their input.

It means shifting primary learning responsibility from the student to the school.

. . . But the new kind of accountability we are developing in Florida is the kind that holds the educational establishment, from top to bottom, accountable for the educational achievement of its clients—those students who come to school well prepared to learn as well as those who come with nothing in their backgrounds to equip them for a successful learning experience. (p. 4-D)

Resources

Books

Education

Knezevich, S. J. (ed.) *Administrative Technology and the School Executive: Applying the Systems Approach to Educational Administration.* Washington, D.C.: American Association of School Administrators, 1969. Describes administrative procedures necessary to achieve educational accountability.

Lessinger, L. M. *Every Kid a Winner: Accountability in Education.* New York: Simon & Schuster, 1970. Complete description of the accountability movement, including extensive sections on performance contracting, which is not discussed in our text.

Articles

Education

Chapman, Ruth. "Who's Accountable?", and Christian, Floyd T. "How Accountability Should Work." *St. Petersburg Times*, April 30, 1972, section 4-D. Pros and cons of educational accountability as seen by researchers, teachers, and the State Department of Education in Florida.

Deterline, William A. "Applied Accountability"; Garvue, Robert J. "Accountability: Comments and Questions"; Lessinger, Leon M. "Robbing of Dr. Peter to 'Pay Paul': Accounting for Our Stewardship of Public Education." *Educational Technology*, 11:1–63, 1971. Entire issue is devoted to educational accountability.

Physical Education

Caldwell, Stratton F. "Toward a Humanistic Physical Education." *Journal of Health, Physical Education, and Recreation*, 43:31–32, 1972.

PART THREE
Applications

Part One of this book set the stage for the formal presentation in Part Two of the systems approach to the teaching of physical education. Part Three, which is intended as a resource, describes various applications of this approach.

Chapter Thirteen describes how a programmed instruction text on scoring in wrestling was designed according to the systems approach presented in this text. Included in this chapter are many of the procedures we have described—instructional analysis, identification of entry behaviors, formulation of performance objectives, and criterion-referenced evaluation. The chapter also indicates how a cognitive task in physical education may be easily and effectively presented in the form of a programmed instruction text.

We have stressed throughout the text that the systems approach model presented is not "the" systems approach model. There are many such models, and Chapter Fourteen describes the use of one of them in designing an instructional unit. The two models are compared and contrasted. This chapter can serve as a starting point for investigating other models which may be equally interesting or effective.

Chapter Fifteen, the final chapter, deals with instruction in tennis. It was written by a student who read only Parts One and Two of the text and then attempted, on her own, to apply the principles described in those sections to the design of an instructional unit. She received only minimal assistance from the authors. Her chapter serves as an example of the strengths and weaknesses of this approach to instruction.

13

Wrestling: Developing a Programmed Scoring Text

Many of the concepts presently associated with the systems approach to teaching were first utilized with programmed instruction (PI). The authors' first attempt to use the concepts presented in this text was in a course on PI at Florida State University. The students were instructed at the beginning of the course that they should choose an instructional goal which would lend itself to the major features of programmed instruction. Thus when they reached the part of the model dealing with media selection, they were required to develop a brief programmed text and received additional instruction on this methodology. This chapter contains a slightly edited report written by one of the students in that class.

The subject matter with which this PI text deals, scoring in wrestling, makes it of obvious relevance to this particular book. However, it was also chosen because it highlights several other aspects of the application of the systems approach. The author states that he made repeated attempts to revise components of his text in the light of succeeding ones. For example, he revised the task analysis several times as he attempted to write meaningful performance objectives. In the one area in which he failed to revise the instruction, his students failed to achieve the performance goal he had stated.

This chapter also exemplifies the role students can play in the preparatory stages of an instructional activity. It describes how an experienced author–wrestler determined the content and flow of an early version of his text, how novices helped in the early field testing to identify obvious problems, and how the author used this information to revise and clarify a number of points.

The comments of the students who participated in the final evaluation of the program indicate the effectiveness such a program can have when it is used in a relatively little-known activity area.

Dr. Charles Sproles, Florida State University, was primarily responsible for writing this chapter.

Chapter Contents

Introduction and Problem Identification
Instructional Analysis
Entry Behaviors and Target Population
Documentation of the Performance Objectives
An Early Draft
Final Draft
Evaluation
Instructional Strategy
Preliminary Formative Evaluation
One-on-One
Small Group
Final Formative Evaluation
Summary

Student Objectives

The student who understands the material in this chapter should, with the book closed, be able to:

1. Describe, in general terms, the steps taken by the author to produce the final instructional analysis, entry behaviors, and performance objectives
2. Identify the three types of evaluation used by the author and the types of revision which resulted from each
3. Describe the author's rationale for why the users of the programmed instruction text did not achieve the 90/90 standard which he had set

Introduction and Problem Identification

A programmed instruction text should help the teacher present subject material and at the same time enable the student to learn more efficiently on an individual basis. This text is designed to do both with the selected topic.

The topic to be programmed is scoring in wrestling. Many people fail to enjoy or appreciate amateur wrestling because they do not understand the objectives and how the wrestlers may achieve them. In most sports it is obvious to the casual observer who the winner is, but not in wrestling. The confusion is compounded by the renegade "professional rasslin'" shown on television. Thus many people are "turned off" by wrestling before they learn enough about it to appreciate the clean, skillful, demanding competition involved.

Scoring methods in wrestling are very stable, with only minor changes from year to year. The rules and the techniques of scoring are becoming more standardized worldwide as high school, college and Olympic competition becomes more popular.

A text on scoring would aid the physical education teacher in several ways. It would provide individual instruction for every student. It could be used on an alternating basis when lack of ample mat space made it impossible for all class members to be active at the same time. But most importantly, this text would fill a gap in a neglected area. Most physical education instructors are so busy teaching the basic wrestling skills that the concepts of scoring are brushed aside or given only superficial coverage.

Spectators also need to be educated about scoring in wrestling. Often a parent or friend watches a meet in dismay at the things he does not understand, only to ask at the conclusion, "Did you win?"

Testing in this area is simple. A student may be required to score the events of a match in class as they happen. Or he may be given a statement of a hypothetical situation which requires him to respond with the correct score. Ordinary paper-and-pencil tests may also be used.

The terminal objective in behavioral terms was stated as follows:

> Given a statement of events which occur in a wrestling match, the student will assign the proper points to each event and calculate the match score; then, having determined the match winners, he will assign meet points and calculate the meet winner with 90 per cent accuracy.

Instructional Analysis

The instructional analysis for scoring a wrestling match or meet does not involve a great deal of physical activity. Nevertheless, the following subtasks were listed in order to detect every detail of the task. (See Figure 13-1 for the relationships of the subtasks.)

1. The scorer positions himself where he can see the match and the referee.
2. He observes a scoring maneuver on the mat.
3. He hears the referee confirm the scoring maneuver.
4. He records the maneuver on the score sheet.
5. He assigns the correct number of points to said maneuver.
6. He observes that the match is over by looking at the clock or by being told by the referee. If the match ends with a fall, the scorer observes the elapsed time and records it with the winner's name.
7. He adds up the match points and records the total, including possible riding time.
8. He tells the referee who the winner is.
9. He writes down the proper number of meet points.
10. He adds meet points to accumulating totals for the teams.
11. He indicates the correct meet score in scorebook or on public scoreboard.

"*Most . . . are so busy teaching the basic wrestling skills that the concepts of scoring are brushed aside or given only superficial coverage.*"

12. He is ready to observe the next match.
13. He declares the meet winner at the end of the meet.

The author was the subject matter expert for the programmed text. He used as a reference *The Official National Collegiate Athletic Association Wrestling Guide* (Cranbury, N.J.: Barnes, 1971). His interpretations of scoring as outlined in this guide were aided by ten years' experience as a wrestling coach and official. The author consulted with someone more familiar with programming in order to determine the extent to which he should cover his subject in this text. It was decided that the task would be limited to assigning points and manipulating scoring factors rather than teaching the student or spectator about the theory or performance of skills.

After several revisions, a final instructional analysis was made. The major revision of the instructional analysis consisted of including in the entry behavior a knowledge of what a takedown (or any other manuever) involved. It was also decided that if the program were applied to spectators who did not meet this behavioral criterion, the referee's announcements of events would serve as ample stimuli. The instructional analysis was further revised by deleting the special scoring systems used in tournament wrestling, which are less practical for the physical education class or the spectator at a dual meet.

FIG. 13-1 *Task analysis for scoring a wrestling match*

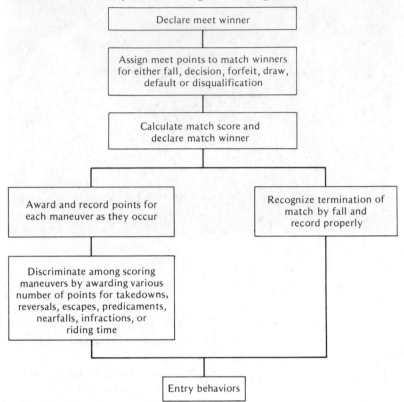

Entry Behaviors and Target Population

The individual who takes this course must possess certain qualities and meet minimum qualifications.

1. He must be able to read and write on a fifth-grade level.
2. He must be able to add whole numbers to three digits.
3. He must be able to work independently.
4. A prior knowledge of wrestling terminology would be desirable so that he could recognize a scoring maneuver. (This behavior would not be required in all cases.)

The target population includes students in beginning wrestling classes, at either the junior high, high school, or college level. They will have acquired some entry behaviors in class, which will provide them with additional assurance of success in this program.

There is a secondary target population, which includes parents and other spectators who possess the necessary entry skills except for an understanding of the terminology. The referee could remedy this deficiency by informing the spectators every time a pertinent scoring maneuver occurred.

This dual target population seems reasonable to the author and seems to further justify his program. Since no beginning wrestling classes were available, the testing population was a group of spectators from the secondary target population, who were interested in wrestling but had no intimate knowledge of it.

Documentation of the Performance Objectives

Performance objectives state exactly what the student is expected to do under given conditions and how well he is expected to perform. The most difficult part of writing performance objectives was to discern and define some nearly covert acts. It seems that one must learn to think in terms of performance objectives before he can precisely describe a student's observable actions. The author had to reexamine what he meant by some terms such as "to score," meaning to write down the points rather than just know how many a maneuver was worth. Over and over again it was necessary to translate into behavioral terms such phrases as "to know" or "to understand." The final objectives bear little resemblance to the first draft.

After changing the terminal objective to include meet scoring, it was necessary to write additional performance objectives at the top of the hierarchy. Because it would have made the program too long, the recognition-discrimination-by-observation phase was converted to an entry behavior and was thus deleted from the list of performance objectives. The tournament scoring tasks and their accompanying objectives were also eliminated.

The standards of performance were revised downward, since they might not have been realistic for the beginner or spectator for whom the program is intended. The author felt that the objectives could and should be met with 100 per cent accuracy.

The final draft of the performance objectives was written with direct reference to the final instructional analysis, beginning with the lowest step in the hierarchical structure and progressing through the final task. Care was taken to include important conditions, behavioral terms, and an efficiency criterion for each objective.

Early Draft

1. Given the scoring maneuver, the student will be able to ascribe the appropriate number of points and tally the score with 100 per cent accuracy.
2. Given a list of scoring situations in any match, the student will be able to determine the winner in 100 per cent of the cases.
3. Whenever a fall is recorded, the student will be able to point out the winner with 100 per cent accuracy and properly record the results.

4. Given the accumulated riding time, the student will determine if a match point has been earned and award it to the proper wrestler with 90 per cent accuracy.

5. Given a match in progress, the student will recognize on sight any scoring maneuver and record it with 90 per cent accuracy.

Final Draft

1. Given a list of scoring maneuvers, the student will ascribe the appropriate number of points to each maneuver with 90 per cent accuracy.

2. Given the accumulated riding time, the student will determine if a match point has been earned and award it to the proper wrestler with 90 per cent accuracy.

3. Given a list of maneuvers in any match, the student will assign the appropriate number of points to each maneuver and calculate the match score with 90 per cent accuracy.

4. When a fall occurs, the student will recognize the termination of the match and record the results with 90 per cent accuracy.

5. Given the termination of a match, the student will tally the final match score and declare the winner with 90 per cent accuracy.

6. Given the results of any number of matches, the student will ascribe the proper number of meet points to each match with 90 per cent accuracy.

7. Given the results of all the matches in a meet, the student will calculate the meet score and winner with 90 per cent accuracy.

Evaluation

The pretest/posttest was designed to measure the student's knowledge of the items involved in the performance objectives. It was also keyed to the instructional analysis. It was a straightforward test in which the student was asked to define, list, or supply basic information. An effort was made to keep it simple and to relate it directly to the tasks in the analysis. Repetition was avoided, and all important objectives were covered. Only a limited number of the test questions, such as Question 1, require something other than exact response; these items involve key words. The pretest and posttest appeared to be valid, reliable, and objective. The difference between the pre- and posttest scores should provide a clear indication of how much learning occurred. The test is reproduced below.

There are 40 responses to this test, and each one has equal value.

1. Define (and make a clear distinction between):
   a. match scoring _____
   b. meet scoring _____

"What is the maximum number of penalty points that can be awarded to a wrestler before his opponent is disqualified?"

2. List seven different ways to gain points in match scoring and the number of points awarded for each maneuver:

   Maneuver   *Points*

   _____   _____
   _____   _____
   _____   _____
   _____   _____
   _____   _____
   _____   _____
   _____   _____

3. How many times may a wrestler be penalized for infractions before being disqualified? _____
4. What is the maximum number of penalty points that can be awarded to a wrestler before his opponent is disqualified? _____
5. What is the largest margin by which a wrestler may win a match? _____

6. List the terms used to describe the outcomes of matches and the point value of each for meet scoring.

Terms   *Points*

7. What is the maximum number of points that may be scored in a high school meet _____; college meet _____?
8. Riding time may earn a wrestler _____ point(s) if he accumulates _____ more minute(s) than his opponent.
9. When a fall occurs, it means that a match is _____, and the score is recorded as _____ – _____.

   Circle *True* or *False* in numbers 10 and 11.

10. *True* or *False*   It is possible to win more than one half of the matches and lose the meet.
11. *True* or *False*   A high-scoring, one-sided decision is the best way to win a match.

Instructional Strategy

This program was prepared in the manner in which the author felt that it would be most serviceable. It can be used with beginning wrestling classes, as was the original intention, but it can also be just as effective with the general population, male or female. The author decided upon a linear program because he felt that it was the most efficient for the purposes outlined in the performance objectives. The subject matter did not seem diverse enough to require or benefit from a branching program. The linear program allows enough practice for the slower student but not so much that the beginner becomes impatient. An effort was made to keep the responses brief in order to hold the student's attention and yet not bore him. The strategy was to progress from short, simple frames to longer, more complex ones with varying modes of response.

Several types of frames were used in this text. The majority are of the constructed-response type, which require very few overt responses. Nevertheless, the author felt that an overt response was desirable and that short ones seem to encourage the student, especially when used successfully early in a program.

Some discrimination-type frames were used, usually as a follow-up to several concepts which had been mentioned earlier. This type of frame seems especially effective for relating ideas which have been introduced previously and for adding variety to the program.

Some copying frames were used but they seem justified because of the nature of the material. These frames have the same effect as extra practice when terms must be precisely defined, as was the case in parts of this paper.

The program was first written on 4″ x 6″ cards with the answers on the back. Cards could be and were added and/or removed during the early evaluation. The program was then printed on ditto sheets which were cut in half, leaving one frame per sheet for group testing. These results will be discussed in a later section. The program was then revised for the last time and produced in the following form.

Samples from Programmed Instruction Text on Scoring in Wrestling	You may begin by placing your name in the following spaces: _____ , _____ last name first Complete each frame by placing your answer in the space indicated, or follow the directions given. Begin.
Frame 1	There are several ways to score points in a wrestling match. One 2-point method is to score a takedown. A takedown is worth _____ points. Answer: 2
Frame 2	There are several other ways to score 2 points in wrestling. A nearfall can also be worth the same number of points as a takedown. A nearfall can therefore be worth _____ points. Answer: 2
Frame 3	A wrestler who has been taken down may even the score by making a reversal. A reversal is also worth _____ points. Answer: 2
Frame 4	Check the scoring maneuvers which may be worth 2 points each. _____ reversal _____ takedown _____ escape _____ nearfall _____ pin Answer: reversal, takedown, nearfall

Frame 5

Two-point maneuvers are the most popular ways of scoring in wrestling. List three maneuvers which are worth 2 points each.

_____

_____

_____

Answer: reversal, takedown, nearfall

Frame 6

A wrestler who is in a position of disadvantage may get himself to a neutral position and thus earn 1 point for an escape. An escape is not worth as much as a takedown, nearfall, or reversal, but is still a desirable maneuver when _____ point(s) is/are needed.

Answer: 1

. . . .

Frame 31

A decision may be worth either 3 or 4 team points. A decision by 9 match points or less is worth 3 points in meet scoring. Therefore, if wrestler *A* wins his match by a 5–4 decision, his team will be given _____ points.

Answer: 3

Frame 32

A decision won by 10 match points or more is worth 4 meet points to the winning team. A decision by a match score of 18–1 is more valuable to the winning team than a 3–2 decision. Thus from a team standpoint it is to the winner's advantage to secure the 4-point decision or a fall. A fall is worth 3 more points than a 3-point decision and _____ more points than a _____.

Answer: 2, 4-point decision.

Frame 33

As was mentioned earlier, a college meet consists of ten matches. If for any reason one team fails to have a representative in a given weight, then the representative of the other team is declared the winner by forfeit. A forfeit is of the same meet value as a fall, default or disqualification; it therefore has a value of _____ meet points.

Answer: 5

Frame 34

Mark an *x* by the three groups of match results which have equal value in *meet* scoring.

Group 1 __	Group 2 __	Group 3 __	Group 4 __
3-pt. decision	fall	fall	fall
3-pt. decision	default	fall	default
4-pt. decision	forfeit	4-pt. decision	disqualification
4-pt. decision		3-pt. decision	
4-pt. decision			

Answer: 1, 2, 4

Frame 35

As a quick review of meet scoring, place the correct number of meet points by the six possible outcomes of a match.

_____ 1. draw   _____ 4. default
_____ 2. fall   _____ 5. disqualification
_____ 3. forfeit   _____ 6. decision

Answer: 1. 2   4. 5
          2. 5   5. 5
          3. 5   6. 3

Frame 36

The team which wins four matches by falls will beat the team which wins the other six matches by decisions. The score will be _____–
_____.

Answer: 20–18

Frame 37

The team that wins four matches by falls will beat the team which wins the other six matches by 3-point decisions. The score will be

_____–_____.

Answer: 24–18.

Frame 38

It is good for a wrestler to win his match by a large margin (18–2); but it is better for him to win by a _____.

Answer: Fall

Congratulations, you have completed the programmed instruction. You are now ready to "show off" what you have learned. See your instructor about the posttest.

Preliminary Formative Evaluation

One-on-One

The formative stage of evaluation included a series of one-on-one operations. The author sat with the subject and took notes as the subject progressed through the program. In order to determine the overall continuity, clarity, and appropriateness of the test, the author first gave it to a person who knew wrestling.

The result of this test indicated that the content was accurate

"It is good for a wrestler to win his match by a large margin . . . but it is better for him to win by a ...."

but that some of the frames were poorly written and needed rewording. One frame was considered too long and was cut down and reworded. The subject had no trouble with the content or concepts and completed the program in about 22 minutes.

The next phase of the one-on-one testing involved similar sessions with two novice subjects, who had only a casual acquaintance with wrestling and could be considered in the same category as beginning students or spectators. Both subjects were students at Florida State University. The results of these evaluations were combined for expediency.

1. The subjects took longer to complete the program (30–40 minutes).
2. Their mistakes seemed to be due to misreading and poor wording.
3. They had trouble with identical frames, especially Frame 18.

Revisions were made after the two novice sessions of one-on-one evaluation as follows:

1. The wording of several frames was changed slightly to eliminate ambiguities.
2. Frame 18 was replaced by a completely new frame incorporating a different approach to the same idea. The author felt that the idea was important and should remain in that position in the program.
3. The mechanics of two frames were changed to assure that the proper blank would be related to the correct group of terms.
4. Two new frames were added following Frame 22. It was felt that some additional reinforcement was needed on the importance of recording fall time as part of the score.

5. The last frame was added as a summation frame, with no overt response called for.

Small Group

The small-group evaluation was planned and carried out with five people who met the entry qualifications but lacked experience in or knowledge of wrestling. These criteria were confirmed by the pretest.

Finding people who knew very little about wrestling was not a problem. Three women were included in the small-group evaluation. A programming expert was also included, and his comments and suggestions were of particular value. The subjects were pretested, given the program, and posttested in one sitting. They were told that this was an evaluation of the program and were asked to offer suggestions. They were invited to write their comments directly on the frames if they so desired.

The main purpose in this phase was to see how much the subjects had learned from the program and to time the three-phase process. It was necessary to test the subjects in two groups, but the conditions were the same for each group, and there were no interactions between the author and the subjects during the testing. After the testing the author interviewed each subject briefly and made notes which he used in subsequent revisions. The results of the small-group evaluation follow.

Results of the Testing

1. The subjects improved their scores by a large amount. The raw score increased from a mean of 6 correct answers on the pretest to a mean of 33, out of a possible 40. This indicates that the correct answers increased from 12 per cent on the pretest to 81 percent on the posttest.
2. The mean time of the program was 35 minutes. The longest time required was 50 minutes, and the shortest 24 minutes.
3. Frame error rate was not calculated at this time but was observed to be very small.

Suggestions by Subjects

1. Frames 5 and 14 might be unnecessary because of repetition.
2. A different response should be used in Frame 16.
3. Frame 18 might need elaboration.
4. Frames 20 and 22 might be too easy. (The subject missed the objective of Frame 22 on the posttest.)
5. The authenticity of the rule in Frame 21 should be checked.
6. Frame 31 has two possible answers.
7. Some review frames might be needed near the end. (The subjects could not remember some items covered early in the program.

Revisions after Small-group Evaluation

1. Frame 14 was reworded and changed to a true-false statement.
2. A second response was added to Frame 18 for clarity.
3. In Frame 20, which was deemed too easy, the author included the figuring of match scores rather than just pointing out ties.
4. Frame 21 was revised to include a statement about normal occurrence.
5. In Frame 29 the two-letter cues (fa, de, etc.) were removed from response blanks as unnecessary prompts.
6. A completely new frame was added as an extra review.
7. Many other minor revisions were also made as a result of the small-group evaluation. It appears to have been the most valuable of the evaluations, perhaps because one of the subjects was experienced in programming.

Final Formative Evaluation

The final formative evaluation called for the testing of a group of ten subjects, who were taken from the secondary target population since no beginning wrestling classes were available at the time.

A minor limitation was placed on the summative evaluation in that the subjects were not tested as a group. The author felt that this limitation could be compensated for by the fact that all subjects were tested under the same conditions and according to written instructions. Any teacher could have administered the program guided by these instructions, which are reproduced below.

Instructions to the administrator:

• This program, including pretest and posttest, may be given in one normal class period.
• Students should be encouraged to guess on the pretest, as a wrong answer will not count against them.
• All answers will go into spaces provided on the tests and program.
• First distribute the pretest, then give directions orally concerning name, date, etc. Before students begin, give them instructions for the program and also the posttest.

Program instructions:

1. Place your name in the proper place on the program.
2. Place the answer in the space provided on each frame.

3. Turn the page and check your answer. If you missed it, do not go back and change it. Just make sure you understand the correct answer.

4. When you have completed the program, indicate that you are ready for the posttest by. . . .

Posttest instructions:

1. Complete the posttest in the same manner as the pretest.
2. Return it to the instructor when you have finished.

A test population was selected which could meet the entry qualifications. It included both males and females; most were students between the ages of 18 and 25. The group could be considered a random selection, with no special learning problems.

The pretest, program, and posttest were to be given at one sitting, and the time spent on each segment was to be observed. All tests were scored by the author, and the data tabulated. A portion of the results of the formative evaluation results follows.

The error rates on all frames were low. Table 13-1 shows the frame error rate, but after the errors were tabulated, the table seemed unnecessary. Errors involved only 10 per cent of the subjects per frame except for Frame 18, on which 20 per cent made errors. More than 60 per cent of the subjects made no errors.

The test results indicate that the program does teach what it is supposed to teach. The 90/90 standard was almost achieved. All

TABLE 13-1 *Summative test results*

Subjects	Per cent correct			Time		
	Pretest	*Posttest*	*Gain pre-post*	*Pre*	*Program*	*Post*
ED	12.5	80.	67.5	5	24	12
NJ	15.	85.	70.	6	50	10
LL	7.5	72.5	65.	4	47	10
MD	2.5	75.	72.5	5	40	11
JL	5.	92.5	87.5	4	35	12
BH	2.5	90.	87.5	4	37	15
TG	7.5	85.	77.5	5	40	10
JA	12.5	87.5	75.	6	35	10
DC	10.	90.	80.	6	32	13
RJ	7.5	77.5	70.	5	25	8
$\overline{X}$	8.25	83.5	75.25	5	36.5	11.1

the subjects exceeded the 70 per cent achievement level, as the posttest indicates. The author feels that the nature of Test Item No. 2 is of poor quality and thus was responsible for holding down the posttest scores. Item No. 2 involves total recall, which is difficult for persons who cannot visualize events as a beginning wrestler would be able to. Thus the author feels that a group in the prime target population would have some advantage over the secondary group on the posttest.

The subjects' attitude was good; they were very receptive to the programmed instruction. Several commented that they felt they had "really" learned something that was new to them. Some seemed fascinated by the uniqueness of the experience. One programming expert termed the instruction "very interesting," while others expressed a desire to see the finished product.

Summary

This programmed text was written to meet the needs of both teacher and student. Limited testing indicated that it could do both.

An analysis of the posttest indicated that Items 2 and 5 were responsible for 96 per cent of the mistakes and that they would have to be revised. The test was the only part of this project which was not continuously revised.

There seemed to be no end to the process of revision and documentation, as is characteristic of the systems approach. The program is therefore incomplete.

14

Weight Training:
A Different Systems Approach

You may be wondering whether systems approach methods other than the model proposed in this book can be effectively applied to physical education. Henry Lehmann (1968) has devised such a method, which we will apply to the teaching of weight training. Weight training was chosen because it is increasingly popular among high school and college students, both as a sport in itself and as a supplement to other sports. The chapter compares and contrasts the model developed by Lehmann with the one presented in this text.

Chapter Contents

Student Objectives

The student who understands the material in this chapter should, with the book closed, be able to:

1. Discuss the similarities and differences between Lehmann's systems approach and the one proposed in this book
2. Describe the rationale for media selection for teaching a specific weight-training skill
3. Apply Lehmann's steps in teaching a weight-training unit

Dr. Ellington Darden, Florida State University, was primarily responsible for writing this chapter.

Comparison of Systems

As you know from previous chapters, the systems approach applied to physical education activities involves an integration of detailed analysis, synthesis, modeling, equipment, media, method, and personnel performing the functions required to accomplish the objectives in the most efficient manner. Lehmann's approach includes the following steps:

1. Need: The educational/training problem
2. Objectives: Measurable learning goals
3. Constraints: Restrictions or limitations
4. Alternatives: Candidate solutions
5. Selection: Choice of best alternative
6. Implementation: Description and operation of the chosen solution
7. Evaluation: Measurement of results obtained against originally stated objectives.

Lehmann's model and the text model both begin by defining a need or problem. The text model is more specific in that it points out the importance of instructional goals or terminal objectives. Lehmann next lists behavioral objectives, while we believe that prior to defining them a thorough investigation should be made of the task as well as students' entry skills. However, Lehmann does mention certain restrictions or limitations after stating his objectives rather than beforehand. The text suggests that evaluative criteria be described after the performance objectives are listed, while Lehmann discusses this step later, when he calls for a number of candidate solutions from which the best alternative is to be selected. Furthermore, Lehmann combines our Steps 6, 7, and 8 (instructional sequence, media selection, and instructional materials) under the heading "implementation." However, both models are consistent in indicating that the systems approach needs to be frequently evaluated and modified to develop maximum student performance.

In reviewing these comparisons, you should recognize that our model tends to be more process-oriented; we believe that instructors should have specific and thorough knowledge of the task and their students. On the other hand, Lehmann tends to emphasize trial-and-error combinations of alternatives which have been effective in the past. We will briefly discuss the application of his system to weight training, demonstrating how each step leads to a more effective program.

Lehmann's Model

Lehmann's first step is to state the real need of the group under consideration, i.e., the overall educational training problem to be solved. This step is but a small part of what this text terms the iden-

Need

tification of instructional goals (see block one in the text model). If we examine some of the current weight-training literature and practices, we immediately note that most of the training systems are slow, boring, nonprogressive, and unrelated to the individual's needs and wants. Therefore, within the context of this problem, the need seems to be for a fast, effective weight-training unit that can be applied within a high school physical education program.

Objectives

Now that the need is properly recognized, descriptions of terminal objectives must be stated in order to meet it. Students should be aware of the specific measurable behaviors they should be able to accomplish after completing the learning experience. From the early chapters you will recall that the formation of performance objectives is the most important step in the systems approach, since all subsequent steps are designed to enable students to achieve those objectives. In the text model the performance objectives are preceded by a complete examination of the instructional goals and the entry skills. Lehmann discusses these processes in his next step. With this step in mind, the following objectives are suggested for possible use in a weight-training unit:

1. The student will be able to demonstrate his muscular strength and endurance by performing the following exercises in competitive form at least once or as many repetitions as possible.
   A. Press (60 percent of body weight)
   B. Bench press (80 percent)
   C. Squat (100 percent)
   D. Curl (50 percent)
   E. Chin (no added resistance)
   F. Sit-up (weight held behind neck) (20 percent)
2. The student will be able to analyze photos taken of himself as well as of other students before and after the weight-training unit, noting improvements in muscular size and shape according to a check list.
3. The student will display interest and a positive attitude by keeping a neat, up-to-date record card.
4. The student will be able to describe and correctly perform weight-training exercises for the major muscles of the body.
5. The student will be able to demonstrate a knowledge of weight-lifting, elementary physiology, and elementary anatomy by correctly answering at least 60 percent of the questions on a written examination.

Constraints

Before we can survey the possible solutions to our problem, we must discuss what Lehmann terms constraints. It is important for you to remember that all solutions have certain constraints or limitations. These constraints may or may not be changed, depending on whether they are governed by laws of nature, fiscal limitations, or tradition. Our weight-training unit will be restricted for the most part to male high school students who have had very little or no experience in lifting weights; however, these students should be neither extremely overweight nor very thin, as the training of either type of individual must be closely supervised. Perhaps you have often observed how these students are ridiculed by others and become discouraged very quickly. For more specific information on underweight and overweight individuals, you may want to examine various height and weight charts (e.g., Wohl and Goodhart, 1968, p. 10). Generally speaking, anyone who weighs less or more than the given range for his height should be placed in a special class.

Other limitations which merit consideration concern time, space, and equipment. In all probability this beginning class will meet in three 50-minute class periods a week for ten weeks. This means that the exercises must be arranged to allow the student to work the major muscles (thighs, back, calves, waist, chest, arms, shoulders, forearms, and neck) of the body in 40 minutes or less. A room approximately 30' × 90', well-ventilated and with sturdy reinforced wooden floors, will adequately accommodate from 20 to 30 men.

Naturally, the weight-training program will be limited to the equipment available. The following equipment is suggested for an adequate training room.

1. 8 6-foot bars with inside and outside collars
2. 4 1-foot dumbbell bars with collars
3. 2000 pounds of assorted weight plates, ranging from 1¼ to 50 pounds.
4. 1 sturdy squat rack
5. 2 benches with attached bench press standards
6. 2 flat benches
7. 1 abdominal or sit-up board
8. 1 chinning bar
9. 1 set of parallel bars

Alternatives

From Lehmann's point of view alternative solutions or approaches should be gathered from a wide spectrum of sources. In the text model alternatives would be included under Step 6, instructional sequence and strategy. The popular "brainstorm" approach often produces alternatives. Therefore, a thorough review of the literature is necessary. Examples of four such alternatives or approaches to weight-training programs are presented for your examination.

The Ross System This system was designed by Charles Ross (Salvati, 1965) for developing maximum muscular gains with a minimum of effort. According to this system the subject performs one repetition with a moderately heavy weight; and when the muscle contraction reaches its peak, he holds it at maximum tension for five seconds. He returns the weight to the floor, takes about five deep breaths, and repeats the procedure about ten times. He then follows the same regimen with another exercise. Ross recommends the following exercises as being most effective in his system:

1. Tricep extension
2. Rowing motion
3. Bicep curl
4. Stiff-legged deadlift
5. Front squat
6. Bench press

The Goldenberg System Joe Goldenberg (1961) believes that a muscle developed using heavy weights with few repetitions lasts only a short time; therefore, he says, every exercise should consist of three sets of 10 to 15 repetitions performed with lighter weights. One should never work until exhausted, but only until pleasantly tired. He recommends that these exercises be performed in the following order:

1. Straddle hop
2. Arm curl
3. Breathing squat
4. Press behind neck
5. Jumping squat
6. Deadlift
7. Calf raise
8. Sit-up
9. Leg raise

The Peripheral Heart Action System This system was developed by Bob Gajda (1966), a former Mr. America winner. The aim of the P.H.A. system is continuous circulation, keeping the blood moving in and out of the muscles at all times. The subject moves constantly. He never needs to stop for rest because he never exercises the same muscle groups in succession, and thus he avoids the exhaustion and congestion induced by many other systems.

The P.H.A. system is divided into several sequences. A sequence is a group of from four to six different exercises, each one for a different part of the body. The student rotates within the sequence for the required number of sets and repetitions. Anywhere from three to six different sequences can be performed in a workout. It is recommended that most sequences consist of from two to three sets for six to ten repetitions, with no rest between exercises. Two recommended sequences are as follows:

Sequence 1	*Sequence 2*
1. Press behind neck	1. Calf raise
2. Calf raise on one leg	2. Side lateral
3. Leg raise	3. Sit-up
4. Bench press	4. Parallel bar dip
5. Chin	5. Lateral pull-down
6. Barbell curl	6. Squat

The Jones System   Arthur Jones (1970) advocates the performance of from six to twelve strict repetitions of a given exercise, followed by partial repetitions or movements until the bar almost falls from the hands. For best results his routine must be followed to the point of momentary exhaustion. In order to aim at that state, the subject must often be "pushed" or coached past his normal limits. However, individuals willing to train this intensely have achieved phenomenal results in short periods of time. Jones recommends the following exercises for a beginner:

1. Squat
2. Press
3. Chin
4. Bench press
5. Parallel bar dip
6. Curl
7. Wrist curl
8. Stiff-legged deadlift
9. Calf raise on one leg
10. Sit-up with bent knees

It is important for you to remember that these alternatives are just *four* of the many training approaches that you should review before selecting the best system. Be as objective as possible in generating a list of alternatives, and don't be afraid to list radical solutions or ideas that at first seem impractical or inappropriate. They may, in the long run, turn out to be your best suggestions.

Selection

To select the best alternative, evaluate all of them systematically, being as open-minded as possible. Selection would fall under Step 6 in the text model. First, go back to your previously stated objectives and constraints, and then list your selection criteria. The following examples are presented as guidelines for helping you develop criteria appropriate to your particular situation.

1. The weight-training program is restricted to male high school students who have had little or no weight-training experience.
2. There will be from 20 to 30 males in a class that meets for 50 minutes Monday, Wednesday, and Friday for ten weeks, in an area 30' x 90'.
3. A minimum but adequate amount of weight-training equipment will be available.
4. The chosen system should work the major muscles of the body in 40 minutes or less.
5. The system should develop strength in the following specific exercises: press, bench press, squat, curl, chin, and sit-up.
6. The weight-training system should develop noticeable improvements in muscle size and shape.
7. The system should be functional enough to allow for occasional lectures, demonstrations, and testing.

Given these seven criteria, your next procedure is to evaluate the various alternatives. Although various computerized mathematical models have been designed to perform this function in business and science, other, less sophisticated models can be used effectively in educational situations. For example, each alternative can be rated against the selection criteria on a $(+, 0, -)$ or $(0, 1, 2)$ basis. The latter point scale might be designed to serve this purpose:

0 = not applicable
1 = applicable
2 = very applicable

By adding the points of each alternative, you can rank the four previously described weight-training systems according to total points. A close examination of Table 14-1 should help clarify this procedure for you.

After you have narrowed your alternatives down to several choices, a more extensive examination will be necessary. Perhaps you need additional opinions and suggestions.

Table 14-1 indicates that the P.H.A. and Jones weight-training

TABLE 14-1 *Evaluation of weight-training systems*

Criteria	Weight-training systems			
	Ross	Goldenberg	P.H.A.	Jones
Beginning students	0	1	2	2
Apply to classes (20–30 students)	0	1	2	2
Alternate day	2	2	2	2
40-minute workout for major muscles	0	0	2	1
Minimum equipment	2	0	2	2
Strength development in 6 lifts	0	0	1	1
Improvements in size and shape	1	1	1	2
Functional	1	2	2	2
Total	6	7	14	14

0 = not applicable
1 = applicable
2 = very applicable

approaches are much more suitable for our needs than the Ross and Goldenberg approaches. The big differences between the first two systems seem to be in the ordering of exercises and intensity of training. Jones believes one should work the largest muscles first and the smaller ones last, always performing each exercise to the point of exhaustion. The P.H.A., on the other hand, does not stress these points. However, Jones does note that in a class of 20 to 30 students, it would be impractical as far as time and equipment are concerned for everyone to begin training by working the largest muscle groups first. Perhaps the best of both systems can be used quite successfully by taking twelve carefully selected exercises and develop three sequences of four exercises, as follows:

Sequence 1	Sequence 2	Sequence 3
1. Squat	1. Stiff-legged deadlift	1. Bench press
2. Pullover	2. Chin	2. Curl
3. Calf raise	3. Press	3. Wrist curl
4. Sit-up	4. Neck bridge	4. Parallel bar dip

Implementation

Lehmann's systems approach is implemented by adopting the selected alternative to meet the specified objectives. Our model tends to simplify the concept of implementation by dividing it into instructional sequence and strategy, media selection, and instructional materials. However, Lehmann recommends that the program be well planned and that it include such items as grouping, scheduling events, and resource requirements. Several factors in the implementation of the weight-training unit warrant discussion, namely:

initial class meetings, grouping, the layout of the room, and the description and illustration of the chosen exercises.

You should use one of the first class periods to discuss such topics as history, misconceptions, and principles of weight training. For example, assigned and recommended readings could be selected from the writings of Jones (1970), Massey et al. (1959), O'Shea (1969), and Sills et al. (1962). After the first week of class an occasional lecture might be necessary to clarify any problems that arise, but lectures should be kept to a minimum.

The grouping of the students at each sequence is an important aspect of successful training. With a maximum of twenty-four students in the weight-training class, there will be eight at each of the three sequences. Within each sequence the students alternate back and forth through each exercise until the sequence is completed or the time has elapsed. It is important that the students at each sequence be of approximately equal strength, so that the changing of weight can be kept to a minimum. It would be ideal from your point of view if all students were of equal size and strength. But this will probably never occur. However, after the first week of testing, you can group the students into a sequence according to general overall strength. Motivation can be kept at a high level if the student is pushed by the students performing in front as well as in back of him. Again, performing each exercise to the point of momentary exhaustion, or until no movement is possible, is a must in order to stimulate maximum growth. However, to avoid muscular soreness, all students should follow a "break-in" program during the first week of training, performing only one set of exercises and terminating them before reaching a point of muscular failure.

For maximum effectiveness students should perform the exercises in close proximity to one another. Figure 14-1 shows a layout for the three described sequences in a 30' x 90' area which may help you organize your school's training room and equipment more efficiently.

You should spend the first week of class explaining the objectives of the course, giving directions for the care of the equipment, demonstrating and explaining the exercises as well as certain safety factors, filling out the weight training record cards, taking Polaroid pictures and making videotapes.

Exercises It is important to explain and demonstrate each exercise and to be sure that the student performs it properly. To help you recognize what these procedures entail, illustrations and descriptions are presented of three of the suggested weight-training exercises and of the major muscles involved and the equipment

FIG. 14-1 *A well-ventilated 30' x 90' weight room*

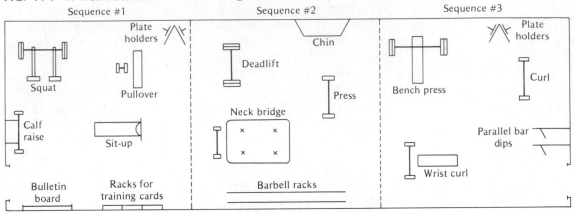

Sequence #1 | Sequence #2 | Sequence #3

Plate holders
Squat
Pullover
Calf raise
Sit-up
Bulletin board
Racks for training cards

Deadlift
Chin
Press
Neck bridge
Barbell racks

Plate holders
Bench press
Curl
Parallel bar dips
Wrist curl

needed. For additional information on the other exercises mentioned in this section, refer to Sills et al. (1962) and O'Shea (1969).

Squat

Major muscles exercised. Prime movers: quadriceps, gluteus maximus, hamstrings, and erector spinae.

Equipment. Squat racks and barbell.

Starting position. In a standing position, hold and support the barbell across the back of the shoulders. The feet should be shoulder-width apart and the back arched, with the head up at all times.

Movement. Lower the body in a controlled manner to a full squat position, and return to the starting position. Take a deep breath and repeat.

Pullover

Major muscles exercised. Prime movers: posterior deltoids, triceps, lower pectoralis major; assistant movers: latissimus dorsi and serratus anterior.

Equipment. Bench and dumbbell.

Starting position. Assume a supine position crosswise on a bench with the shoulders in contact, and head and lower body relaxed and off the bench. Hold a large dumbbell over the chest in a straight-arm position with the thumbs-over grip.

Movement. Taking a deep breath, lower the weight behind the head, emphasizing the stretching of the chest and torso, and return to the starting position.

"Lower the body in a controlled manner to a full squat position, and return to the starting position. Take a deep breath and repeat."

Press

Major muscles exercised. Prime movers: deltoids and triceps; assistant movers: upper trapezius, upper pectoralis major and latissimus dorsi.

Equipment. Barbell.

Starting position. In a standing position, hold the barbell in front of the chest with the hands shoulder-width apart and palms turned away from the body.

Movement. Push the weight upward until the arms are fully extended above the head, then return to the starting position.

Record Card The weight-training record and class card serves as a daily and term record of the student's progress. The front of the card is used to record the student's height and weight as well as his strength (poundage and repetitions) for each of the six listed exercises (see Figure 14-2). These measures are taken at the beginning and end of the quarter. On the back of the card the student keeps a daily training record of all twelve exercises, noting the weights, sets, and repetitions (see Figure 14-3). You can easily check students' progress by glancing through the cards. To simplify roll taking, keep these record cards in a wall rack where students can pick them up at the beginning of each period.

"Push the weight upward until the arms are fully extended above the head, and then to the starting position."

Photos and Videotapes Polaroid pictures should be taken and videotapes made of each student during the first and last weeks of class. For the photos the student should assume a front and back relaxed and flexed double-arm pose. He should be clothed in shorts and photographed against an uncluttered background. Also, it is important to standardize the pose and distance from the camera for all subjects on both occasions. Videotapes can be recorded of the entire class, as well as individuals, as they are actually performing the exercises. The resulting visual feedback can enhance teaching. As a further incentive, and if time and money allow, you can use videotapes to record performances throughout the entire quarter. A student could observe and analyze his performance during class and at specific times outside of class. Videotapes are an excellent means by which you and your students can observe changes that occur in performance during the quarter. Also, the tapes could motivate students by stimulating correct performance.

Evaluation

Two types of evaluation are relevant in applying the systems approach to weight training. First, you must evaluate and grade students according to the objectives you initially specified. Second, you must evaluate the weight-training unit according to its effectiveness in meeting the needs and wants of those involved. Lehmann and we agree that the preceding steps in our models need

FIG. 14-2 *Weight-training record card (front)*

Name _____

Quarter _____   Section _____

Instructor _____

Exercise	1st practical date: _____ Body wt. _____ Ht. _____			2nd practical date: _____ Body wt. _____ Ht. _____		
	Lbs.	Reps.	Pts.	Lbs.	Reps.	Pts.
Bench press (80%)						
Curl (50%)						
Squat (100%)						
Press (60%)						
Sit-up (20%)						
Chin (body wt.)						

Total pts. _____

Average _____   _____

FIG. 14-3 *Weight-training record card (back)*

Exercise	Date: _____		Date: _____		Date: _____		Date: _____		Date: _____	
	wt.	reps.	wt.	reps.	wt.	reps.	wt.	reps.	wt.	reps.
Squat										
Pullover										
Calf raise										
Sit-up										
Deadlift										
Chin										
Press										
Bridge										
Bench press										
Curl										
Wrist curl										
Dips										

frequent evaluation and revision. You must gather and evaluate data to provide feedback for yourself and your students.

Evaluating and Grading the Students The objectives for possible use in our weight-training unit are listed on page 309. The students can be evaluated for the quarter by the following means:

1. Two practical tests, administered in the first and final weeks of the quarter (Objectives 1 and 4).
2. Analyses of physique photos taken during the first and final weeks of the quarter (Objective 2).
3. A neat and up-to-date record card (Objective 3).
4. Written examination based on exercises, lectures, assigned readings, and analyses of exercises on videotapes (Objectives 4 and 5).

Practical Tests The practical test is composed of six basic exercises that are performed with the following percentages of the student's body weight, with as many repetitions as possible:

1. Bench press (80 per cent of body weight)
2. Press (60 per cent)
3. Curl (50 per cent)
4. Squat (100 per cent)
5. Sit-up (20 per cent)
6. Chinning (no added resistance)

Standard scores based on the number of repetitions completed for each of these six exercises have been calculated by O'Shea (1969) in his book *Scientific Principles and Methods of Strength Fitness.* For example, a student who weighs 156 pounds would be required to use 125 pounds (80 percent of his body weight) in the bench press. Using this poundage, he performs five repetitions, which, according to O'Shea, equal 44 points. The student records this number on his record card by writing the poundage, repetitions, and points under the appropriate headings. He then performs and scores the remaining exercises in the same manner. To keep the student from being discouraged, he is also given points when he is unable to perform any repetitions with the required weight. When the student has completed all six exercises (which usually takes two training days), he totals the card and determines the average score. Points given for the average scores, as well as improvement, can be determined by referring to O'Shea's book. For instance, an average of 41 on the first practical test would be equal to 8 points. On the second practical exercise the student might average 56, which would be equal to 18 points. The improvement $(18 - 8 = 10)$

would be worth an additional 12 points. You should note that since the practical test is given during the first and last weeks of the quarter, a change in body weight is likely to occur. Therefore, the poundages should be adjusted accordingly.

Physique Photos The physique photos taken of each student during the first and final weeks of the quarter can be evaluated according to the following numerical check list:

Directions: Place the appropriate number in the space opposite each body part as you observe both photos.

2 = much improvement in muscular development

1 = slight improvement

0 = no improvement

_____ A. Neck

_____ B. Forearm

_____ C. Arms (biceps and triceps)

_____ D. Shoulders

_____ E. Back

_____ F. Chest

_____ G. Waist

_____ H. Thighs

_____ I. Calves

Total points

Each student should score his own before-and-after photos with the approval of several other students. You should double check any questionable improvements. A possible 18 points can be gathered from this scoring system.

Record Card At the end of the quarter you can evaluate the record card on a 10-point scale according to neatness, correctness, and attendance.

Written Examination You will probably want to administer a written examination at the end of the quarter covering the exercises, lectures, assigned readings, and the analyses of the exercises on videotape. A sample test follows.

Final Written Exam in Weight-training

I. True or False:

1. Deep breathing is important in supplying oxygen to the blood stream.

2. Repetitions are of very little importance in a weight-training program.
3. The bench press develops the triceps.
4. The bicep is the largest muscle of the arm.

II. Listing

1. List five reasons for disqualification in an Olympic lifting contest.
2. List five weight-training exercises that work the muscles of the lower body.
3. List 10 major muscles of the upper body.
4. List the power lifts and Olympic lifts.

III. Define

1. Progressive resistance
2. Muscle-bound
3. Cheating
4. Compound exercises

IV. Discussion

1. What are the differences between Olympic lifting and power lifting?
2. Discuss how a winner is determined in a weight-lifting contest.
3. How would you convince a high school boy that weight training would help him?
4. How is a winner determined in a physique contest?

V. Analyses of videotapes

You can show videotapes of various weight-training exercises and have students criticize them according to form and performance.

Grading the Student   The following breakdown of one student's scoring should help you get an overall picture of the suggested grading procedures.

First practical	8
Second practical	18
Improvement	12
Photo evaluation	10
Record card	8
Written exam (50 pts. maximum)	42
Total points	98 points

Scores are then placed on a curve with other beginning classes for final grade distribution if necessary. Or a fixed point total for each letter grade can be established at the outset, and students can be motivated to achieve the number of points required for an "A."

Evaluating the Weight-Training Unit

In order to evaluate the weight-training unit, you must compare the students' performances and behaviors with the initially specified objectives. You can do this in several ways. You can check the class records every three weeks in order to note the students' progress. You can keep a diary in which to record certain strengths and weaknesses of the daily training sessions. For example, you might note that most of the students are incorrectly performing the barbell curl; therefore, you may need to spend more time explaining the proper performance of it in the future.

It is also a good idea to meet with your students every three or four weeks to discuss and evaluate past instructional and training procedures, as well as to plan together future goals for the class. If these meetings result in better communication with your students, particularly with respect to the objectives, then you can use certain criticisms and suggestions to modify the weight-training unit for the benefit of future students. It is also helpful to give the students an opportunity to evaluate the entire weight-training unit as well as the instruction, either on a departmental student evaluation form or in a brief, unsigned paper. As the model indicates, this information will be very valuable to you.

Those objectives not achieved must be carefully studied. No doubt some changes will have to be made after each weight-training unit is completed. These changes will again be implemented, evaluated, and modified, and this iterative process will be continued until the objectives are met to the desired extent.

Summary

In this chapter Lehmann's systems approach was applied to the teaching of a high school weight-training unit. You should be able to discuss the similarities and differences between Lehmann's approach and ours; describe the rationale for the medium selected for teaching a specific weight-training skill; understand and apply Lehmann's steps in teaching a weight-training unit; and evaluate your work in terms of the procedures recommended in this chapter.

It is hoped that in your future role as teacher and coach, you will experience the positive effects of applying various systems approaches to physical education activities, and the satisfaction of teaching students to understand and condition their bodies through weight training.

Resources

Gajda, Bob. "The New Revolutionary P.H.A. or Sequence System of Training." *Iron Man*, 26:14–17, 1966.

Goldenberg, Joseph. "Training with Weights." *Scholastic Coach*, 30:60–63, 1961.

Jones, Arthur. *Nautilus Training Principles, Bulletin No. 1.* DeLand, Fla.: Arthur Jones Productions, 1970.

Lehmann, Henry. "The Systems Approach to Education." *Audiovisual Instruction*, 13:144–148, 1958.

Massey, Benjamin, et al. *The Kinesiology of Weight Lifting.* Dubuque, Iowa: Brown, 1959.

O'Shea, John Patrick. *Scientific Principles and Methods of Strength Fitness.* Reading, Mass.: Addison-Wesley, 1969.

Salvati, Michael J. *The Production of Muscular Bulk.* Alliance, Neb.: Iron Man Industries, 1965.

Sills, Frank D., et al. (eds.). *Weight Training in Sports and Physical Education.* Washington: AAHPER, 1962.

Wohl, Michael G., and Goodhart, Robert S. *Modern Nutrition in Health and Disease.* Philadelphia: Lea & Febiger, 1968.

15

Tennis: Applying
a Systems Model

In this chapter the systems model presented in this book will be used in developing portions of a beginning tennis unit. The application of each step will be described, insofar as it can be expressed in words. Examples of performance objectives from each of the four behavioral domains will be developed. The author is aware that different teachers may approach this unit in different ways, but hopes that it conveys an idea of how the model might be applied.

This section of the book plays a crucial role; the value of the systems approach depends on its applicability to teaching situations.

Chapter Contents

Instructional Goal
Instructional Analysis
Entry Skills and Knowledge
Performance Objectives
Criterion-Referenced Testing
Instructional Sequence and Strategy
Media Selection
Securing Instructional Materials
Formative Evaluation
Summary

Student Objectives

The student who understands this chapter should, with the book closed, be able to:

1. Describe the steps taken to develop a teaching unit using the systems model described in this book
2. Develop a unit on physical education activity using the systems approach

Dr. Barbara Hollingsworth, Florida State University, was primarily responsible for writing this chapter.

325

Instructional Goal   The student should successfully participate in a game of tennis, appropriately observing rules and etiquette.

Instructional Analysis   The instructional analysis is derived from the instructional goal. The question was asked, "What would the learner have to know how to do in order to perform the instructional goal?" The answer is given below in the form of seven prerequisite skills, which are then broken down into subskills. Fig. 15-1 illustrates this process. Only one of the skills, the serve, is used as an example. In order to develop a complete hierarchy, you would have to further analyze each of the skills into subskills.

Be able to stroke the ball effectively.
1. Be able to hold the racket correctly.
2. Be able to anticipate where the ball is heading and how to return it effectively.
3. Be able to use the forehand stroke effectively.
4. Be able to use the backhand stroke effectively.
5. Be able to volley effectively.

Be able to serve effectively.
1. Be able to stand properly in the correct place on the court.
2. Be able to hold the racket properly.
3. Be able to throw the ball up properly.

FIG. 15-1 *A tennis partial learning hierarchy*

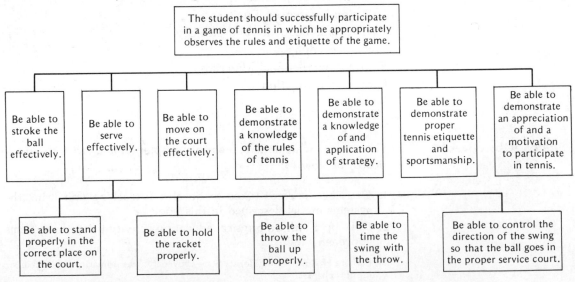

4. Be able to time the swing with the throw.
5. Be able to control the direction of the swing so that the ball goes into the proper service court.

Be able to move effectively on the court.

1. Be able to move with speed and agility.
2. Be able to move and control the body so as to be in position to return serves and volleys.

Be able to demonstrate a knowledge of the rules of the game.

1. Be able to recall and apply the rules of tennis.
2. Be able to keep score correctly.
3. Be able to recall the players' serving order.

Be able to demonstrate knowledge and application of simple strategy.

1. Be able to assume proper court position one or two feet behind the baseline when waiting for opponent's shot.
2. Be able to keep the ball deep.
3. Be able to move the opponent by hitting the ball away from where he is.
4. Be able to play offensively by approaching the net.
5. Be able to defend against an attack by using lobs.

Be able to demonstrate proper tennis etiquette and sportsmanship.

1. Be able to demonstrate proper player conduct.
2. Be able to demonstrate proper spectator conduct.
3. Be able to demonstrate the ability to call balls correctly and honestly.

Be able to demonstrate an appreciation of and a motivation to participate in tennis.

1. Be able to demonstrate positive attitudes toward the game and toward participation.
2. Play outside of class (on own time).
3. Play for personal pleasure and play with enthusiasm.

Entry Skills and Knowledge

It is assumed that this unit is designed for a coed high school class. At this stage of the model, you should assess entry skills and knowledge. Knowledge of rules, etiquette, scoring, etc. can be assessed by a written recall-type test. Skill level can be assessed by pretests. The pretests for the forehand, backhand, and serve will be the same as the skill tests used on the posttest. The students will be asked to fill out a questionnaire with items pertaining to their sex, age, cognitive abilities, personal goal level, motivation, attitudes, interest, and previous experiences in tennis.

Students must have minimum entry abilities, such as hand-eye coordination, intelligence sufficient for learning, etc., in order to stay in the class. These can be assessed by checking I. Q. scores and by simple coordination tests, such as the Crawford Small Parts Dexterity Test. The information revealed by these tests and the questionnaire will be used in determining at what point to begin instruction and in helping to plan general and individual instructional methods and strategies.

Performance Objectives

Two objectives will be selected as examples from each of the four domains: psychomotor, cognitive, affective, and social. These eight objectives, which are derived from the instructional analysis (See Figure 15-2), will then be further analyzed in the next steps of the systems model.

Psychomotor Domain

1. The right-handed student will stand on the right side of a regulation tennis court. The teacher, who is on the left side of the opponent's court, will throw 20 balls overhand to the student. (Reverse for left-handed student.) Using the forehand stroke, the student should be able to hit 15 of the 20 balls back to the same side and within the doubles court.
2. The student will stand in the proper position for serving on a regulation tennis court and serve ten balls from the right side. He should be able to complete, with reasonably correct form, 8 out of 10 attempted serves into the correct service court.

FIG. 15-2   *Derivation of tennis performance objectives*

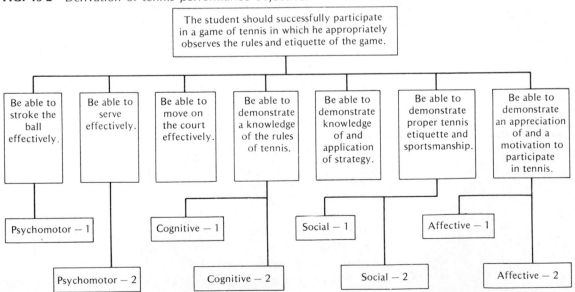

Cognitive Domain

1. The student should be able to observe a singles game in class and keep score correctly with no more than one error.
2. The student should be able to answer correctly 90 percent of the items on a written examination on the rules of tennis. The examination will consist of 20 multiple-choice, 10 completion, and 20 true-false questions.

Affective Domain

1. The student should be motivated to play more tennis during his free time than he did before taking the tennis class.
2. The student should show an interest in tennis by observing one match of the school's varsity tennis team.

Social Domain

1. The student should be able to participate in a tennis singles game with another member of the class and call all balls correctly, as judged by the teacher.
2. In tournament play in class the student should be able to demonstrate emotional stability by controlling his feelings and temper, as judged by the teacher.

Criterion-Referenced Testing

(Note: In this section and hereafter, the objectives will not be written out again but will be referred to by domain and number. For example, Psychomotor—1 refers to the first psychomotor objective, which is concerned with stroking the ball.)

Psychomotor—1. The right-handed student will stand on the right side of the court. The teacher, who is on the left side of the opponent's court, will throw 20 balls to the student. (Reverse for left-handed student.) In order to pass, the student must, using the forehand stroke, hit 15 of the 20 balls back to the same side of the court and within the doubles court.

Psychomotor—2. With the teacher observing, the student will stand in proper serving position and serve 10 balls from the right side of the court. In order to pass, he must hit 8 of the 10 balls into the correct service court.

Cognitive—1. With the teacher observing, the students will keep score on paper of three singles games played in class by two demonstrators from outside the class. The student should make no more than one error.

Cognitive—2. On a written examination of tennis rules, which consists of 50 recall and recognition items, the student must be able to recall 90 per cent of the answers.

Affective—1. The student will participate in tennis games more than he did before taking this course. Whether or not students do this will be revealed by their responses to a questionnaire

". . . the student will stand in proper serving position and serve 10 balls from the right side of the court."

with items indicating how often they participate in tennis, etc.

Affective—2. The student must observe at least one match of the school's varsity tennis team. If the school does not have a team, he must observe another school's team or watch a match on television.

Social—1. With the teacher observing, the student will participate in a tennis game with another player and call all balls correctly, as judged by the teacher.

Social—2. With the teacher observing and judging, the student should be able to play a match in a class tournament and display appropriate emotional control and good sportsmanship.

Instructional Sequence and Strategy

The GAMES and PRACTICE plans will be used in developing the strategy and sequence for the first psychomotor objective.

Psychomotor—1.

Preparing the student (the GAMES approach):

Prepractice Events	*Presentation Modes*
Goal orientation	visual, verbal
The student should be aware of standards against which he will be evaluated; the image of the ideal stroke should be clear in his mind.	Observation of correct execution of forehand drive from a loop film; demonstration and explanation of correct execution of forehand drive; demonstration of where to stand, how to stand, and how to hold racket.
Attention	visual, verbal, kinesthetic
The student should be alert and responsive to instruction and appropriate cues.	The learning environment—number and condition of courts, amount of equipment, teaching techniques, etc.—should stimulate and maintain attention.
Motivation	verbal
The student should be optimally motivated to perform.	Since the task is moderately difficult for most beginners, general incentives to stimulate learning should also be moderate. Externally and internally oriented motivation may be applied where appropriate.
Experienced for transfer	verbal, visual
The student should have experienced prerequisite skills and knowledge that he can transfer in a positive manner to the present situation. Furthermore, he should be informed of these relationships.	For the benefit of students who have played softball, baseball, or badminton, explain and demonstrate similarity between forehand drive and batting or forehand stroke in badminton.
Situation familiarity	visual, verbal
The student should be familiar with the essence of the learning situation, the nature of the general cues to which he will respond. In this case, he will perform on the tennis courts, in response to actions made by his opponent.	Showing and discussing the aspects of moving on tennis court and strategy as to where to move in what situations, when to begin backswing, when to use forehand, etc. The physics and mechanics of stroking the ball should be discussed.

Selecting and sequencing appropriate learning events (the PRACTICE approach):

Practice Events	*Presentation Modes*
Practice	kinesthetic, visual, verbal
The student should practice as frequently as necessary until he has mastered the task. Practice of the task can be self-paced at first, but will be externally paced during a game situation which will increase the difficulty.	At first the student can perform the task by dropping the ball and hitting it. Then he must practice with a partner. The student must be concerned with fast, unpredictable stimuli to which he must respond instantaneously. Overpractice for long-term retention benefits.

Reinforcement

The student should be at least intermittently reinforced as he approaches skill mastery.

verbal

Comments, praise, rewards, grades, other forms of acknowledgment.

Administration

The student should master this task, with the necessary prerequisite skills, in two or three practice sessions. Specific actual task practice should be relatively massed.

kinesthetic, verbal, visual

Although the task involves competing responses, skill should be demonstrated after several practice sessions. If skill is not being attained, rest or other forms of learning may be appropriate.

Contiguity

The student should execute with appropriate timing the series of stimulus-response acts that constitute the chain of events of which the forehand stroke is composed.

kinesthetic, verbal

Individual components of the task are cued by the teacher and practiced in proper spatial and temporal arrangement. Timing is important in this task.

Tradeoffs

The student should practice parts of the task and then the entire task; and he should stress accuracy of stroke over speed; primarily physical instead of mental practice; social (paired) practice instead of individual practice; and real instead of simulated conditions in most cases.

kinesthetic

Correct execution of forehand is more important than fast stroke; task complexity and organization dictate use of the part-method approach; practice is primarily physical, but the student could imagine the correct execution of the strokes outside of class; emphasis on paired practice to improve social interaction; some simulation (ball-throwing machine) used but practice should usually occur under real conditions.

Information feedback

The student should be aware of his skill development as he proceeds in practice.

kinesthetic, verbal, visual

Internal (kinesthetic and visual) sources provide some concurrent information about form, execution, and accuracy. Teacher comments and videotape are external sources of terminal feedback and more precise information about performance.

Cue generalization and discrimination

The student should proceed from general concepts and responses to highly specific acts.

kinesthetic, visual, verbal, auditory

Meaningful and appropriate cues at each phase of learning (words cue and direct the student during the sequence of movements; model films and videotape continually inform him where he is and where he should be; manual manipulation of passive limbs gives him the feel of the movement). Concentration and attention are fairly important during

kinesthetic, visual, verbal, auditory

practice of this task, as it involves the execution of cued responses.

Event repetition

The student should repeat the task frequently according to specific requirements.

kinesthetic

On the tennis courts, where he will be tested and will play games

Alternatives: practice against a backboard. When weather does not permit practice on courts, students may practice in gymnasium.

Because of space limitations, the GAMES and PRACTICE plans for the seven remaining objectives have been synthesized into an instructional sequence and an instructional strategy for achieving each objective.

Psychomotor—2.

Sequence

1. Know where and how to stand
2. Know how to hold the racket
3. Know how to throw the ball up
4. Be able to time the swing with the throw
5. Know how to hit the ball
6. Be able to control the direction of the ball

Strategy

1. Demonstrate and explain the correct method of service (using the part method because of the complexity of the skill).
2. Discuss the mechanical principles involved, e.g., throw the ball high and hit it at an angle so it will go over the net and then down into the service court.
3. For the benefit of students who have played badminton, compare and contrast badminton and tennis serves.
4. Show a loop film on service.
5. Have the students practice serving on the court, alternating with a partner. Since the task is self-paced, students should practice for accuracy and consistency first, and emphasize speed later.
6. Cue individual components of the serve and have students practice them in proper spatial and temporal order.
7. Provide feedback as to how students are progressing in skill development. Students also receive visual feedback by seeing where the ball goes.
8. Use a videotape recorder to provide feedback on occasion.

Cognitive—1.

Sequence

1. Be able to recall the rules of game.
2. Know the numbers and terms used in scoring.
3. Know what constitutes a point.
4. Know what a game score is.
5. Know what match and set scores are.

Strategy

1. Read a tennis rule book.
2. Go over the rules and explain them.
3. Explain how to keep score.
4. Demonstrate how to keep a game score with two students playing an actual game.
5. Have some students keep score while other members of the class play.
6. Have the students play and keep score of their own game.
7. Reinforce students' correct scoring.

Cognitive—2.

Sequence

1. Be familiar with the rules of tennis.
2. Understand the rules.
3. Be able to recall the rules.

Strategy

1. Read a tennis rule book.
2. Go over the rules and explain them. Also explain the need for rules and describe their relationship from sport to sport.
3. Have the students play games in class, observing rules. Point out any infractions of rules.
4. Comment on and praise the students' knowledge of rules.

Affective—1.

Sequence

1. Understand the game.
2. Be able to play a game successfully.
3. Know the strategy of playing tennis.
4. Know when tennis courts are available.
5. Want to participate in tennis.

Strategy

1. Teach the students how to play a game successfully.
2. Teach them the rules of the game.
3. Have them demonstrate that they can play a game successfully in class.
4. Give them written information stating when and where tennis courts are available.
5. Inform them of proper player conduct.
6. Encourage them to participate in intramurals.
7. Reinforce motivation and interest, reward and praise participation.
8. Provide a variety of situations designed to shape positive values.

Affective—2.

Sequence

1. Understand the game.
2. Have good experiences in games played in class.

Strategy

1. Explain and demonstrate the game of tennis.
2. Briefly instruct the students in the strategy of playing tennis.
3. Give them information about the school's tennis team.
4. Instruct them in proper spectator conduct.
5. Encourage them to observe matches.
6. Reinforce their interest.

Social—1.

Sequence

1. Know the rules for calling balls.
2. Find good examples to follow in class.
3. Explain and observe the effects of desirable behaviors on groups.

Strategy

1. Teach students the rules concerning lines and calling balls.
2. Emphasize the importance of calling balls honestly. Explain the relationship between honesty in a sports situation and honesty in other forms of socialization.
3. Set a good example at all times.
4. Reinforce appropriate behavior.

Social—2.

Sequence

1. Know and understand tennis etiquette.
2. Be aware of the importance of observing proper etiquette and conduct.

Strategy

1. Explain proper tennis etiquette and sportsmanship and the relationship of good sportsmanship in tennis to social interactions in real life.
2. Set a good example at all times.
3. Reinforce appropriate behavior.

Media Selection

Psychomotor—1.

1. Lecture on correct execution of forehand drive.
2. Demonstration of forehand drive.
3. Loop film on forehand drive—for modeling behavior.
4. Videotape recorder to provide feedback.
5. Ball-throwing machine for actual practice.

Psychomotor—2.

1. Lecture on service.
2. Demonstration of service as a modeling behavior.
3. Loop film on service to illustrate proper form.
4. Videotape recorder to provide feedback.

Cognitive—1.

1. Tennis rule book.
2. Chalkboard for illustrating how to keep score.
3. Demonstration of keeping score.

Cognitive—2.

1. Tennis rule book.
2. Verbal explanation of rules.
3. Mimeographed handout on rules.

Affective—1.

1. Mimeographed handout on strategy.
2. Mimeographed handout containing all necessary information

about where tennis courts are located and when they are available.

3. Lecture on proper player conduct.

Affective—2.

1. Mimeographed handout on strategy.
2. Newspaper articles concerning tennis team.
3. Lecture on proper spectator conduct.

Social—1.

1. Tennis rule book.
2. Lecture on proper tennis etiquette and sportsmanship.

Social—2.

1. Lecture on proper tennis etiquette and sportsmanship.

Securing Instructional Materials

At this point in the systems model, the instructional materials would be selected, or developed if none were available. It is difficult to successfully explain this process on paper, but, in general, it would consist of the following steps: The first logical step would be to select resource books from which to prepare handouts and lectures. First, select a rule book and, if possible, make it available to students. You might also select a text for students to use. If not, you will have to prepare handouts and lectures from these resource books and from personal experience. The next step would be to visit the nearest media center and select films and filmstrips for the unit. This would also be a good time to reserve projectors and other necessary mechanical devices. Also make plans to reserve ball-throwing machines and the videotape recorder from the department or school, or whoever is in charge of these machines. One final step would be to obtain information from the library, intramural office, coach, etc., concerning tennis intramurals and the school's tennis team.

Formative Evaluation

The first step in the formative evaluation process is to summarize the data collected from the students. Then organize it according to each instructional activity, and evaluate it. You should make a chart containing information on students' mean performance on the pretest and the posttest. For example, the first psychomotor objective pertained to the forehand stroke. The criterion for achieving this objective was correctly hitting 15 of 20 balls. The criterion

". . . demonstration and explanation of correct execution of forehand drive; demonstration of where to stand, how to stand, and how to hold racket."

for the second psychomotor objective was to serve 8 of 10 attempts correctly. The data summary for these two objectives might look something like that in Figure 15-3. These data simply indicate that for the objective relating to the forehand stroke, the average number of correctly stroked balls was 5 on the pretest and 16 on the posttest. Five percent of the class met the criterion of 15 on the pretest and 80 percent reached it on the posttest. Performance relating to every objective would be analyzed in this manner. You should inspect these data carefully in order to evaluate students' performance. From a summary of these data, you should be able to evaluate the effectiveness of the instruction and identify problem areas.

In the content revision process the first items to be reevaluated are the test items. They need to be reviewed in order to determine their validity with regard to their relationship to the objectives. You will need to obtain a rough estimate of the reliability of these items in order to determine whether they consistently measure the achievement of the objectives. You can determine their reliability by examining the relationship among various items which have been used to measure the same objective. If students who got an item right for an objective tended to get other items right, then the instrument would tend to be reliable.

Next you should reexamine the scores on the test of entry behavior to see if the students actually had the skills you assumed they did when they started the instruction. You can obtain this

FIG. 15-3 *Data summary sheet*

	$\overline{X}$	Per cent criterion	$\overline{X}$	Per cent criterion
Psychomotor objective I	5	.05	16	.80
Psychomotor objective II	2	.10	8	.85

information by looking at the summary data chart and checking the percentage of students who had achieved the criterion related to entry behaviors.

Inspect the pretest and posttest data carefully. If the students performed poorly on the pretest and well on the posttest, then you may assume that instruction improved their skills. The posttest data will indicate the students' success or failure in achieving instructional goals.

Looking at pretest and posttest data for the various objectives in the context of the learning hierarchy can be very effective. By assessing the results of the instruction for each subskill, you can pinpoint any weaknesses in the instructional activities.

The next step is to reexamine the performance objectives. You should decide whether the objectives continue to represent the kinds of performance which will be expected from the students.

Another method of formative evaluation is to have students express their attitudes toward the instruction and the content of the course. You could use a written questionnaire and tally the results. Or students could make verbal comments in formal debriefing sessions or in informal moments during class.

After you have collected and examined all the relevant data, you must then revise the instructional content and procedures of the course. You should reexamine each instructional activity to see whether it is consistent with objectives and test items.

The final steps in revision are administering the posttest and assign grades. Time is an important factor in the administration of the posttest. Especially relevant in all instructional situations is the amount of time that elapses before a learning activity is evaluated.

Grading is an important factor. This procedure was described in the section concerning criterion-referenced evaluation. Given a choice, this author would prefer grading based on the attainment of a certain number of performance objectives, indicating the cutoff point. But if letter grades were required, the process would be more complicated. One method would be to preestablish point totals which students would have to achieve in order to obtain various grades. Another method would take into consideration pretest as well as posttest scores. Entry skills, physical characteristics, etc., would be considered. In this writer's opinion, the best method would take into consideration both final performance and improvement. A method which shows promise in this respect is one recently proposed by Hale and Hale,[1] which permits logarithmic conversion of the pretest and posttest scores so that improvement gains from different initial scores can be converted into units for comparison. This formula can be used to determine the rank gain of each student, taking into consideration initial position, amount of gain, and difficulty of the task. Grades would then be assigned on the basis of rank gain.

Summary

In this chapter you have had an opportunity to follow, step by step, the application of the systems approach in developing a teaching unit for tennis. Examples of performance objectives for each of the four behavioral domains were presented and developed. Applications were made in certain cases, and in other cases general procedures were outlined. At this point you should be able to select an activity and develop a unit by stating your instructional goals, analyzing your task, stating your performance objectives, deciding how to evaluate them, developing your instructional sequence and strategy, selecting media, developing or selecting instructional materials, evaluating your unit, and revising it.

[1] Hale, Patricia W., and Hale, Robert M. "Comparison of Student Improvement by Exponential Modification of Test-Retest Scores." *Research Quarterly* 43:113–120, 1972.

Appendix
Behavioral Domains in Detail

Psychomotor Domain

The classification scheme prepared by Elizabeth Simpson for *The Illinois Teacher of Home Economics* (1966–67) is presented here in its entirety. It should provide you with a more complete understanding of psychomotor behaviors and their relation to each other.

1.0 *Perception* This is an essential first step in performing a motor act. It is the process of becoming aware of objects, qualities, or relations by way of the sense organs. It is a necessary but not sufficient condition for motor activity. It is basic in the situation—interpretation—action chain leading to motor activity. The category of perception has been divided into three subcategories indicating three different levels of the perception process. This level is a parallel of the first category, receiving or attending, in the affective domain.

   1.1 *Sensory stimulation*—Impingement of a stimulus or stimuli upon one or more of the sense organs.

   1.11 *Auditory*—Hearing or the sense of organs of hearing.

   1.12 *Visual*—Concerned with the mental pictures or images obtained through the eyes.

   1.13 *Tactile*—Pertaining to the sense of touch.

   1.14 *Taste*—Determine the relish or flavor of by taking a portion into the mouth.

   1.15 *Smell*—To perceive by excitation of the olfactory nerves.

   1.16 *Kinesthetic*—The muscle sense; pertaining to sensitivity from activation of receptors in muscles, tendons, and joints.

The preceding categories are not presented in any special order of importance, although, in Western cultures, the visual cues are said to have dominance, whereas in some cultures, the auditory and tactile cues may pre-empt the high position we give the visual. Probably no sensible ordering of these is possible at this time. It should also be pointed out that "the cues that guide action may change for a particular motor activity as learning progresses (e.g., kinesthetic cues replacing visual cues)."

1.1 *Sensory stimulation*—Illustrative educational objectives.

Sensitivity to auditory cues in playing a musical instrument as a member of a group.

Awareness of difference in "hand" of various fabrics.

Sensitivity to flavors in seasoning food.

1.2 *Cue selection*—Deciding to what cues one must respond in order to satisfy the particular requirements of task performance. This involves identification of the cue or cues and associating them with the task to be performed. It may involve grouping of cues in terms of past experience and knowledge. Cues relevant to the situation are selected as a guide to action; irrelevant cues are ignored or discarded.

1.2 *Cue selection*—Illustrative educational objectives.

Recognition of operating difficulties with machinery through the sound of the machine in operation.

Sensing where the needle should be set in beginning machine stitching.

Recognizing factors to take into account in batting in a softball game.

1.3 *Translation*—Relating of perception to action in performing a motor act. This is the mental process of determining the meaning of the cues received for action. It involves symbolic translation, that is, having an image or being reminded of something, "having an idea," as a result of cues received. It may involve insight which is essential in solving a problem through perceiving the relationships essential to solution. Sensory translation is an aspect of this level. It involves "feedback," that is, knowledge of the effects of the process. Translation is a continuous part of the motor act being performed.

1.3 *Translation*—Illustrative educational objectives.

Ability to relate music to dance form.

Ability to follow a recipe in preparing food.

Knowledge of the "feel" of operating a sewing machine successfully and use of this knowledge as a guide in stitching.

2.0 *Set*—Set is a preparatory adjustment or readiness for a particular kind of action or experience. Three aspects of set have been identified: mental, physical, and emotional.

2.1 *Mental set*—Readiness, in the mental sense, to perform a certain motor act. This involves, as prerequisite, the level of perception and its subcategories. Discrimination, that is, using judgment in making distinctions, is an aspect of mental set.

2.1 *Mental set*—Illustrative educational objectives.

Knowledge of steps in setting the table.

Knowledge of tools appropriate to performance of various sewing operations.

2.2 *Physical set*—Readiness in the sense of having made the anatomical adjustments necessary for a motor act to be performed. Readiness, in the physical sense, involves receptor set, that is, sensory attending, or focusing the attention of the needed sensory organs and postural set, or positioning of the body.

2.2 *Physical set*—Illustrative educational objectives.

Achievement of bodily stance preparatory to bowling.

Positioning of hands preparatory to typing.

2.3 *Emotional set*—Readiness in terms of attitudes favorable to the motor acts taking place. Willingness to respond is implied.

2.3 *Emotional set*—Illustrative educational objectives.

Disposition to perform sewing machine operation to best of ability.

Desire to operate a production drill press with skill.

3.0 *Guided response*—This is an early step in the development of skill. Emphasis here is upon the abilities which are components of the more complex skill. Guided response is the overt behavioral act of an individual under the guidance of the instructor or in response to self-evaluation where the student has a model or criteria against which he can judge his performance. Prerequisite to performance of the act are readiness to respond, in terms of set to produce the overt behavioral act and selection of the appropriate response. Selection of response may be defined as deciding what response must be made in order to satisfy the requirements of task performance. There appear to be two major subcategories, imitation and trial and error.

3.1 *Imitation*—Imitation is the execution of an act as a direct response to the perception of another person performing the act.

3.1 *Imitation*—Illustrative educational objectives.

Imitation of the process of stay-stitching the curved neck edge of a bodice.

Performing a dance step as demonstrated.

Debeaking a chick in the manner demonstrated.

3.2 *Trial and error*—Trying various responses, usually with some rationale for each response, until an appropriate response is achieved. The appropriate response is one which meets the requirements of task performance, that is, "gets the job done" or does it more efficiently. This level may be defined as multiple-response learning in which the proper response is selected out of varied behavior, possibly through the influence of reward and punishment.

3.2 *Trial and error*—Illustrative educational objectives.

Discovering the most efficient method of ironing a blouse through trial of various procedures.

Determining the sequence for cleaning a room through trial of several patterns.

4.0 *Mechanism*—Learned response has become habitual. At this level, the learner has achieved a certain confidence and degree of proficiency in the performance of the act. The act is a part of his repertoire of possible responses to stimuli and the demands of situations where the response is an appropriate one. The response may be more complex than at the preceding level; it may involve some patterning in carrying out the task.

4.0 *Mechanism*—Illustrative educational objectives.

Ability to perform a hand-hemming operation.

Ability to mix ingredients for butter cake.

Ability to pollinate an oat flower.

5.0 *Complex overt response*—At this level, the individual can perform a motor act that is considered complex because of the movement pattern required. At this level, skill has been attained. The act can be carried out smoothly and efficiently, that is, with minimum expenditure of time and energy. There are two subcategories: resolution of uncertainty and automatic performance.

5.1 *Resolution of uncertainty*—The act is performed without hesitation of the individual to get a mental picture of task sequence. That is, he knows the sequence required and so proceeds with confidence. The act is here defined as complex in nature.

5.1 *Resolution of uncertainty*—Illustrative educational objectives.

Skill in operating a milling machine.

Skill in setting up and operating a production band saw.

Skill in laying a pattern on fabric and cutting out a garment.

5.2 *Automatic performance*—At this level, the individual can perform a finely coordinated motor skill with a great deal of ease and muscle control.

5.2 *Automatic performance*—Illustrative educational objectives.

Skill in performing basic steps of national folk dances.

Skill in tailoring a suit.

Skill in performing on the violin.

6.0 *Adaptation*—Altering motor activities to meet the demands of new problematic situations requiring a physical response.

6.0 Adaptation Illustrative educational objectives.

Developing a modern dance composition through adapting known abilities and skills in dance.

7.0 Origination—Creating new motor acts or ways of manipulating materials out of understandings, abilities, and skills developed in the psychomotor area.

7.0 Origination—Illustrative educational objectives.

Creation of a modern dance.

Creation of a new game requiring psychomotor response (pp. 135–141).

Cognitive Domain

When Benjamin Bloom and his co-workers developed their *Taxonomy of Educational Objectives, Handbook I: Cognitive Domain* in 1956, they little knew the influence they would ultimately have on educational practice and policy. A condensed version of their hierarchical organizational plan follows. It's taken from the *Taxonomy of Educational Objectives, Handbook II: The Affective Domain* (1964), by David Krathwohl, Benjamin Bloom, and Bertram Masia.

Knowledge

1.00 *Knowledge*

Knowledge, as defined here, involves the recall of specifics and universals, the recall of methods and processes, or the recall of a pattern, structure, or setting. For measurement purposes, the recall situation involves little more than bringing to mind the appropriate material. Although some alteration of the material may be required, this is a relatively minor part of the task. The knowledge objectives emphasize most the psychological processes of remembering. The process of relating is also involved in that a knowledge test situation requires the organization and reorganization of a problem such that it will furnish the appropriate signals and cues for the information and knowledge the individual possesses. To use an analogy, if one thinks of the mind as a file, the problem in a knowledge test situation is that of finding in the problem or task the appropriate signals, cues, and clues which will most effectively bring out whatever knowledge is filed or stored.

1.10 Knowledge of Specifics

The recall of specific and isolable bits of information. The emphasis is on symbols with concrete referents. This material, which is at a very low level of abstraction, may be thought of as the elements from which more complex and abstract forms of knowledge are built.

1.11 Knowledge of Terminology

Knowledge of the referents for specific symbols (verbal and non-verbal). This may include knowledge of the most generally accepted symbol referent, knowledge of the variety of symbols which may be used for a single referent, or knowledge of the referent most appropriate to a given use of a symbol.

To define technical terms by giving their attributes, properties, or relations. Familiarity with a large number of words in their common range of meanings.

1.12 Knowledge of Specific Facts

Knowledge of dates, events, persons, places, etc. This may include very precise and specific information such as the specific date or exact magnitude of a phenomenon. It may also include approximate or relative information such as an approximate time period or the general order of magnitude of a phenomenon.

The recall of major facts about particular cultures.
The possession of a minimum knowledge about the organisms studied in the laboratory.

1.20 Knowledge of Ways and Means of Dealing with Specifics

Knowledge of the ways of organizing, studying, judging, and criticizing. This includes the methods of inquiry, the chronological sequences, and the standards of judgment within a field as well as the patterns of organization through which the areas of the fields themselves are determined and internally organized. This knowledge is at an intermediate level of abstraction between specific knowledge on the one hand and knowledge of universals on the other. It does not so much demand the activity of the student in using the materials as it does a more passive awareness of their nature.

1.21 Knowledge of Conventions

Knowledge of characteristic ways of treating and presenting ideas and phenomena. For purposes of communication and consistency, workers in a field employ usages, styles, practices, and forms which best suit their purposes and/or which appear to suit best the phenomena with which they deal. It should be recognized that although these forms and conventions are likely to be set up on arbitrary, accidental, or authoritative bases, they are retained because of the general agreement or concurrence of individuals concerned with the subject, phenomena, or problem.

Familiarity with the forms and conventions of the major types of works; e.g., verse, plays, scientific papers, etc.
To make pupils conscious of correct form and usage in speech and writing.

1.22 Knowledge of Trends and Sequences

Knowledge of the processes, directions, and movements of phenomena with respect to time.

Understanding of the continuity and development of American culture as exemplified in American life.
Knowledge of the basic trends underlying the development of public assistance programs.

1.23 Knowledge of Classifications and Categories

Knowledge of the classes, sets, divisions, and arrangements which are regarded as fundamental for a given subject field, purpose, argument, or problem.

To recognize the area encompassed by various kinds of problems or materials.
Becoming familiar with a range of types of literature.

1.24 Knowledge of Criteria

Knowledge of the criteria by which facts, principles, opinions, and conduct are tested or judged.

Familiarity with criteria for judgment appropriate to the type of work and the purpose for which it is read.
Knowledge of criteria for the evaluation of recreational activities.

1.25 Knowledge of Methodology

Knowledge of the methods of inquiry, techniques, and procedures employed in a particular subject field as well as those employed in investigating particular problems and phenomena. The emphasis here is on the individual's knowledge of the method rather than his ability to use the method.

Knowledge of scientific methods for evaluating health concepts.
The student shall know the methods of attack relevant to the kinds of problems of concern to the social sciences.

1.30 Knowledge of the Universals and Abstractions in a Field

Knowledge of the major schemes and patterns by which phenomena and ideas are organized. These are the large structures, theories, and generalizations which dominate a subject field or which are quite generally used in studying phenomena or solving problems. These are at the highest levels of abstraction and complexity.

1.31 Knowledge of Principles and Generalizations

Knowledge of particular abstractions which summarize observations of phenomena. These are the abstractions which are of value in explaining, describing, predicting, or in determining the most appropriate and relevant action or direction to be taken.

Knowledge of the important principles by which our experience with biological phenomena is summarized.
The recall of major generalizations about particular cultures.

1.32 Knowledge of Theories and Structures

Knowledge of the *body* of principles and generalizations together with their interrelations which present a clear, rounded, and systematic view of a complex phenomenon, problem, or field. These are the most abstract formulations, and they can be used to show the interrelation and organization of a great range of specifics.

The recall of major theories about particular cultures.
Knowledge of a relatively complete formulation of the theory of evolution.

Intellectual Abilities and Skills

Abilities and skills refer to organized modes of operation and generalized techniques for dealing with materials and problems. The materials and problems may be of such a nature that little or no specialized and technical information is required. Such information as is required can be assumed to be part of the individual's general fund of knowledge. Other problems may require specialized and technical information at a rather high level such that specific knowledge and skill in dealing with the problem and the materials are required. The abilities and skills objectives emphasize the mental processes of organizing and reorganizing material to achieve a particular purpose. The materials may be given or remembered.

2.00 *Comprehension*

This represents the lowest level of understanding. It refers to a type of understanding or apprehension such that the individual knows what is being communicated and can make use of the material or idea being communicated without necessarily relating it to other material or seeing its fullest implications.

2.10 Translation

Comprehension as evidenced by the care and accuracy with which the communication is paraphrased or rendered from one language or form of communication to another. Translation is judged on the basis of faithfulness and accuracy; that is, on the extent to which the material in the original communication is preserved although the form of the communication has been altered.

The ability to understand nonliteral statements (metaphor, symbolism, irony, exaggeration).
Skill in translating mathematical verbal material into symbolic statements and vice versa.

2.20 Interpretation

The explanation or summarization of a communication. Whereas translation involves an objective part-for-part rendering of a communication, interpretation involves a reordering, rearrangement, or new view of the material.

The ability to grasp the thought of the work as a whole at any desired level of generality.
The ability to interpret various types of social data.

2.30 Extrapolation

The extension of trends or tendencies beyond the given data to determine implications, consequences, corollaries, effects, etc., which are in accordance with the conditions described in the original communication.

The ability to deal with the conclusions of a work in terms of the immediate inference made from the explicit statements.
Skill in predicting continuation of trends.

3.00 *Application*

The use of abstractions in particular and concrete situations. The abstractions may be in the form of general ideas, rules of procedures, or generalized methods. The abstractions may also be technical principles, ideas, and theories which must be remembered and applied.

Application to the phenomena discussed in one paper of the scientific terms or concepts used in other papers.
The ability to predict the probable effect of a change in a factor on a biological situation previously at equilibrium.

4.00 *Analysis*

The breakdown of a communication into its constituent elements or parts such that the relative hierachy of ideas is made clear and/or the relations between the ideas expressed are made explicit. Such analyses are intended to clarify the communication, to indicate how the communication is organized, and the way in which it manages to convey its effects, as well as its basis and arrangement.

4.10 Analysis of Elements

Identification of the elements included in a communication.

The ability to recognize unstated assumptions.
Skill in distinguishing facts from hypotheses.

4.20 Analysis of Relationships

The connections and interactions between elements and parts of a communication.

Ability to check the consistency of hypotheses with given information and assumptions.
Skill in comprehending the interrelationships among the ideas in a passage.

4.30 Analysis of Organizational Principles

The organization, systematic arrangement, and structure which hold the communication together. This includes the "explicit" as well as "implicit" structure. It includes the bases, necessary arrangement, and mechanics which make the communication a unit.

The ability to recognize form and pattern in literary or artistic works as a means of understanding their meaning.
Ability to recognize the general techniques used in persuasive materials, such as advertising, propaganda, etc.

5.00 *Synthesis*

The putting together of elements and parts so as to form a whole. This involves the process of working with pieces, parts, elements, etc., and arranging and combining them in such a way as to constitute a pattern or structure not clearly there before.

5.10 Production of a Unique Communication

The development of a communication in which the writer or speaker attempts to convey ideas, feelings, and/or experiences to others.

Skill in writing, using an excellent organization of ideas and statements.
Ability to tell a personal experience effectively.

5.20 Production of a Plan, or Proposed Set of Operations

The development of a plan of work or the proposal of a plan of operations. The plan should satisfy requirements of the task which may be given to the student or which he may develop for himself.

Ability to propose ways of testing hypotheses.
Ability to plan a unit of instruction for a particular teaching situation.

5.30 Derivation of a Set of Abstract Relations

The development of a set of abstract relations either to classify or explain particular data or phenomena, or the deduction of propositions and relations from a set of basic propositions or symbolic representations.

Ability to formulate appropriate hypotheses based upon an analysis of factors involved, and to modify such hypotheses in the light of new factors and considerations.
Ability to make mathematical discoveries and generalizations.

6.00 *Evaluation*

Judgments about the value of material and methods for given purposes. Quantitative and qualitative judgments about the extent to which material and methods satisfy criteria. Use of a standard of appraisal. The criteria may be those determined by the student or those which are given to him.

6.10 Judgments in Terms of Internal Evidence

Evaluation of the accuracy of a communication from such evidence as logical accuracy, consistency, and other internal criteria.

Judging by internal standards, the ability to assess general probability of accuracy in reporting facts from the care given to exactness of statement, documentation, proof, etc.

The ability to indicate logical fallacies in arguments.

6.20 Judgments in Terms of External Criteria

Evaluation of material with reference to selected or remembered criteria.

The comparison of major theories, generalizations, and facts about particular cultures.

Judging by external standards, the ability to compare a work with the highest known standards in its field—especially with other works of recognized excellence. (pp. 188–193)

Affective Domain

Since the outline of the classification scheme created by David Krathwohl, Benjamin Bloom, and Bertram Masia presented in Chapter Five may be insufficient to describe their work, the following longer but condensed version is presented.

1.0 *Receiving (Attending)*

At this level we are concerned that the learner be sensitized to the existence of certain phenomena and stimuli; that is, that he be willing to receive or to attend to them. This is clearly the first and crucial step if the learner is to be properly oriented to learn what the teacher intends that he will. To indicate that this is the bottom rung of the ladder, however, is not at all to imply that the teacher is starting *de novo*. Because of previous experience (formal or informal), the student brings to each situation a point of view or set which may facilitate or hinder his recognition of the phenomena to which the teacher is trying to sensitize him.

The category of *Receiving* has been divided into three subcategories to indicate three different levels of attending to phenomena. While the division points between the subcategories are arbitrary, the subcategories do represent a continuum. From an extremely passive position or role on the part of the learner, where the sole responsibility for the evocation of the behavior rests with the teacher—that is, the responsibility rests with him for "capturing" the student's attention—the continuum extends to a point at which the learner directs his attention, at least at a semiconscious level, toward the preferred stimuli.

1.1 Awareness

Awareness is almost a cognitive behavior. But unlike *Knowledge*, the lowest level of the cognitive domain, we are not so much concerned with a memory of, or ability to recall, an item or fact as we are that, given appropriate opportunity, the learner will merely be conscious of something—that he take into account a situation, phenomenon, object, or stage of affairs. Like *Knowledge* it does not imply an assessment of the qualities or nature of the stimulus, but unlike *Knowledge* it does not necessarily imply attention. There can be simple awareness without specific discrimination or recognition of the objective characteristics of the object, even though these characteristics must be deemed to have an effect. The individual may not be able to verbalize the aspects of the stimulus which cause the awareness.

Develops awareness of aesthetic factors in dress, furnishings, architecture, city design, good art, and the like.
Develops some consciousness of color, form, arrangement, and design in the objects and structures around him and in descriptive or symbolic representations of people, things, and situations.[1]

1.2 Willingness to Receive

In this category we have come a step up the ladder but are still dealing with what appears to be cognitive behavior. At a minimum level, we are here describing the behavior of being willing to tolerate a given stimulus, not to avoid it. Like *Awareness*, it involves a neutrality or suspended judgment toward the stimulus. At this level of the continuum the teacher is not concerned that the student seek it out, nor even, perhaps, that in an environment crowded with many other stimuli the learner will necessarily attend to the stimulus. Rather, at worst, given the opportunity to attend in a field with relatively few competing stimuli, the learner is not actively seeking to avoid it. At best, he is willing to take notice of the phenomenon and give it his attention.

Attends (carefully) when others speak—in direct conversation, on the telephone, in audiences.
Appreciation (tolerance) of cultural patterns exhibited by individuals from other groups—religious, social, political, economic, national, etc.
Increase in sensitivity to human need and pressing social problems.

1.3 Controlled or Selected Attention

At a somewhat higher level we are concerned with a new phenomenon, the differentiation of a given stimulus into figure and ground at a conscious or perhaps semiconscious level—the differentiation of aspects of a stimulus which is perceived as clearly marked off from adjacent impressions. The perception is still without tension or assessment, and the student may not know the technical terms or symbols with which to describe it correctly or precisely to others. In some instances it may refer not so much to the selectivity of attention as to the control of attention, so that when certain stimuli are present they will be attended to. There is an element of the learner's controlling the attention here, so that the favored stimulus is selected and attended to despite competing and distracting stimuli:

Listens to much with some discrimination as to its mood and meaning and

[1] Illustrative objectives selected from the literature follow the description of each subcategory.

with some recognition of the contributions of various musical elements and instruments to the total effect.

Alertness toward human values and judgments on life as they are recorded in literature.

2.0 *Responding*

At this level we are concerned with responses which go beyond merely attending to the phenomenon. The student is sufficiently motivated that he is not just 1.2 *Willing to attend*, but perhaps it is correct to say that he is actively attending. As a first stage in a "learning by doing" process the student is committing himself in some small measure to the phenomena involved. This is a very low level of commitment, and we would not say at this level that this was "a value of his" or that he had "such and such an attitude." These terms belong to the next higher level that we describe. But we could say that he is doing something with or about the phenomenon besides merely perceiving it, as would be true at the next level below this of 1.3 *Controlled or selected attention*.

This is the category that many teachers will find best describes their "interest" objectives. Most commonly we use the term to indicate the desire that a child become sufficiently involved in or committed to a subject, phenomenon, or activity that he will seek it out and gain satisfaction from working with it or engaging in it.

2.1 Acquiescence in Responding

We might use the word "obedience" or "compliance" to describe this behavior. As both of these terms indicate, there is a passiveness so far as the initiation of the behavior is concerned, and the stimulus calling for this behavior is not subtle. Compliance is perhaps a better term than obedience, since there is more of the element of reaction to a suggestion and less of the implication of resistance or yielding unwillingly. The student makes the response, but he has not fully accepted the necessity for doing so.

Willingness to comply with health regulations.
Obeys the playground regulations.

2.2 Willingness to Respond

The key to this level is in the term "willingness," with its implication of capacity for voluntary activity. There is the implication that the learner is sufficiently committed to exhibiting the behavior that he does so not just because of a fear of punishment, but "on his own" or voluntarily. It may help to note that the element of resistance or of yielding unwillingly, which is possibly present at the previous level, is here replaced with consent or proceeding from one's own choice.

Acquaints himself with significant current issues in international, political, social, and economic affairs through voluntary reading and discussion.
Acceptance of responsibility for his own health and for the protection of the health of others.

2.3 Satisfaction in Response

The additional element in the step beyond the *Willingness to respond* level, the consent, the assent to responding, or the voluntary response, is that the behavior is accompanied by a feeling of satisfaction, an emotional response, generally of pleasure, zest, or enjoyment. The location of this category in the hierarchy has given us a great deal of difficulty. Just where in the process of internalization the attachment of an emotional response, kick, or thrill to a behavior occurs has been hard to determine. For that matter there is some uncertainty as to whether the level of internalization at which it occurs may not depend on the particular behavior. We have even questioned whether it should be a category. If our structure is to be a hierarchy, then each category should include the behavior in the next level below it. The emotional component appears gradually through the range of internalization categories. The attempt to specify a given position in the hierarchy as *the* one at which the emotional component is added is doomed to failure.

The category is arbitrarily placed at this point in the hierarchy where it seems to appear most frequently and where it is cited as or appears to be an important component of the objectives at this level on the continuum. The category's inclusion at this point serves the pragmatic purpose of reminding us of the presence of the emotional component and its value in the building of affective behaviors. But it should not be thought of as appearing and occurring at this one point in the continuum and thus destroying the hierarchy which we are attempting to build.

Enjoyment of self-expression in music and in arts and crafts as another means of personal enrichment.
Finds pleasure in reading for recreation.
Takes pleasure in conversing with many different kinds of people.

3.0 *Valuing*

This is the only category headed by a term which is in common use in the expression of objectives by teachers. Further, it is employed in its usual sense: that a thing, phenomenon, or behavior has worth. This abstract concept of worth is in part a result of the individual's own valuing or assessment, but it is much more a social product that has been slowly internalized or accepted and has come to be used by the student as his own criterion of worth.

Behavior categorized at this level is sufficiently consistent and stable to have taken on the characteristics of a belief or an attitude. The learner displays this behavior with sufficient consistency in appropriate situations that he comes to be perceived as holding a value. At this level, we are not concerned with the relationships among values but rather with the internalization of a set of specified, ideal values. Viewed from another standpoint, the objectives classified here are the prime stuff from which the conscience of the individual is developed into active control of behavior.

This category will be found appropriate for many objectives that use the term "attitude" (as well as, of course, "value").

An important element of behavior characterized by *Valuing* is that it is motivated, not by the desire to comply or obey, but by the individual's commitment to the underlying value guiding the behavior.

3.1 Acceptance of a Value

At this level we are concerned with the ascribing of worth to a phenomenon, behavior, object, etc. The term "belief," which is defined as "the emotional acceptance of a proposition or doctrine upon what one implicitly considers adequate ground" (English and English, 1958, p. 64), describes quite well what may be thought of as the dominant characteristic here. Beliefs have varying degrees of certitude. At this lowest level of *Valuing* we are concerned with the lowest levels of certainty; that is, there is more of a readiness to re-evaluate one's position than at the higher levels. It is a position that is somewhat tentative.

One of the distinguishing characteristics of this behavior is consistency of response to the class of objects, phenomena, etc. with which the belief or attitude is identified. It is consistent enough so that the person is perceived by others as holding the belief or value. At the level we are describing here, he is both sufficiently consistent that others can identify the value, and sufficiently committed that he is willing to be so identified.

Continuing desire to develop the ability to speak and write effectively.
Grows in his sense of kinship with human beings of all nations.

3.2 Preference for a Value

The provision for this subdivision arose out of a feeling that there were objectives that expressed a level of internalization between the mere acceptance of a value and commitment or conviction in the usual connotation of deep involvement in an area. Behavior at this level implies not just the acceptance of a value to the point of being willing to be identified with it, but the individual is sufficiently committed to the value to pursue it, to seek it out, to want it:

Assumes responsibility for drawing reticent members of a group into conversation.
Deliberately examines a variety of viewpoints on controversial issues with a view to forming opinions about them.
Actively participates in arranging for the showing of contemporary artistic efforts.

3.3 Commitment

Belief at this level involves a high degree of certainty. The ideas of "conviction" and "certainty beyond a shadow of a doubt" help to convey further the level of behavior intended. In some instances this may border on faith, in the sense of it being a firm emotional acceptance of a belief upon admittedly nonrational grounds. Loyalty to a position, group, or cause would also be classified here.

The person who displays behavior at this level is clearly perceived as holding the value. He acts to further the thing valued in some way, to extend the possibility of his developing it, to deepen his involvement with it and with the things representing it. He tries to convince others and seeks converts to his cause. There is a tension here which needs to be satisfied; action is the result of an aroused need or drive. There is a real motivation to act out the behavior.

Devotion to those ideas and ideals which are the foundations of democracy. Faith in the power of reason and in methods of experiment and discussion.

4.0 *Organization*

As the learner successively internalizes values, he encounters situations for which more than one value is relevant. Thus necessity arises for (*a*) the organization of the values into a system, (*b*) the determination of the interrelationships among them, and (*c*) the establishment of the dominant and pervasive ones. Such a system is built gradually, subject to change as new values are incorporated. This category is intended as the proper classification for objectives which describe the beginnings of the building of a value system. It is subdivided into two levels, since a prerequisite to interrelating is the conceptualization of the value in a form which permits organization. *Conceptualization* forms the first subdivision in the organization process, *Organization of a value system* the second.

While the order of the two subcategories seems appropriate enough with reference to one another, it is not so certain that 4.1 *Conceptualization of a value* is properly placed as the next level above 3.3 *Commitment*. Conceptualization undoubtedly begins at an earlier level for some objectives. Like 2.3 *Satisfaction in response*, it is doubtful that a single completely satisfactory location for this category can be found. Positioning it before 4.2 *Organization of a value system* appropriately indicates a prerequisite of such a system. It also calls attention to a component of affective growth that occurs at least by this point on the continuum but may begin earlier.

4.1 Conceptualization of a Value

In the previous category, 3.0 *Valuing*, we noted that consistency and stability are integral characteristics of the particular value or belief. At this level (4.1) the quality of abstraction or conceptualization is added. This permits the individual to see how the value relates to those that he already holds or to new ones that he is coming to hold.

Conceptualization will be abstract, and in this sense it will be symbolic. But the symbols need not be verbal symbols. Whether conceptualization first appears at this point on the affective continuum is a moot point, as noted above.

Attempts to identify the characteristics of an art object which he admires.
Forms judgments as to the responsibility of society for conserving human and
material resources.

4.2 Organization of a Value System

Objectives properly classified here are those which require the learner
to bring together a complex of values, possibly disparate values, and
to bring these into an ordered relationship with one another. Ideally,
the ordered relationship will be one which is harmonious and internally
consistent. This is, of course, the goal of such objectives, which seek to
have the student formulate a philosophy of life. In actuality, the integration may be something less than entirely harmonious. More likely
the relationship is better described as a kind of dynamic equilibrium
which is, in part, dependent upon those portions of the environment
which are salient at any point in time. In many instances the organization of values may result in their synthesis into a new value or value
complex of a higher order.

Weighs alternative social policies and practices against the standards of the public
welfare rather than the advantage of specialized and narrow interest groups.
Develops a plan for regulating his rest in accordance with the demands of his
activities.

5.0 *Characterization by a Value or Value Complex*

At this level of internalization the values already have a place in the
individual's value hierarchy, are organized into some kind of internally
consistent system, have controlled the behavior of the individual for a
sufficient time that he has adapted to behaving this way; and an evocation of the behavior no longer arouses emotion or affect except when the
individual is threatened or challenged.

The individual acts consistently in accordance with the values he has
internalized at this level, and our concern is to indicate two things:
(*a*) the generalization of this control to so much of the individual's behavior that he is described and characterized as a person by these pervasive controlling tendencies, and (*b*) the integration of these beliefs,
ideas, and attitudes into a total philosophy or world view. These two
aspects constitute the subcategories.

5.1 Generalized Set

The generalized set is that which gives an internal consistency to the
system of attitudes and values at any particular moment. It is selective
responding at a very high level. It is sometimes spoken of as a determining tendency, an orientation toward phenomena, or a predisposition to
act in a certain way. The generalized set is a response to highly generalized phenomena. It is a persistent and consistent response to a family of
related situations or objects. It may often be an unconscious set which
guides action without conscious forethought. The generalized set may
be thought of as closely related to the idea of an attitude cluster, where
the commonality is based on behavioral characteristics rather than the

subject or object of the attitude. A generalized set is a basic orientation which enables the individual to reduce and order the complex world about him and to act consistently and effectively in it.

Readiness to revise judgments and to change behavior in the light of evidence. Judges problems and issues in terms of situations, issues, purposes, and consequences involved rather than in terms of fixed, dogmatic precepts or emotionally wishful thinking.

5.2 Characterization

This, the peak of the internalization process, includes those objectives which are broadest with respect both to the phenomena covered and to the range of behavior which they comprise. Thus, here are found those objectives which concern one's view of the universe, one's philosophy of life, one's *Weltanschauung*—a value system having as its object the whole of what is known or knowable.

Objectives categorized here are more than generalized sets in the sense that they involve a greater inclusiveness and, within the group of attitudes, behaviors, beliefs, or ideas, an emphasis on internal consistency. Though this internal consistency may not always be exhibited behaviorally by the students toward whom the objective is directed, since we are categorizing teachers' objectives, this consistency feature will always be a component of *Characterization* objectives.

As the title of the category implies, these objectives are so encompassing that they tend to characterize the individual almost completely. (pp. 178–185)

Develops for regulation of one's personal and civic life a code of behavior based on ethical principles consistent with democratic ideals.
Develops a consistent philosophy of life.

Index